Advances in Elastomers

Advances in Elastomers

Editor

Michal Sedlačík

MDPI • Basel • Beijing • Wuhan • Barcelona • Belgrade • Manchester • Tokyo • Cluj • Tianjin

Editor
Michal Sedlačík
Tomas Bata University in Zlín
Czech Republic

Editorial Office
MDPI
St. Alban-Anlage 66
4052 Basel, Switzerland

This is a reprint of articles from the Special Issue published online in the open access journal *Materials* (ISSN 1996-1944) (available at: https://www.mdpi.com/journal/materials/special_issues/adv_elastomers).

For citation purposes, cite each article independently as indicated on the article page online and as indicated below:

LastName, A.A.; LastName, B.B.; LastName, C.C. Article Title. *Journal Name* **Year**, *Volume Number*, Page Range.

ISBN 978-3-0365-0434-6 (Hbk)
ISBN 978-3-0365-0435-3 (PDF)

© 2021 by the authors. Articles in this book are Open Access and distributed under the Creative Commons Attribution (CC BY) license, which allows users to download, copy and build upon published articles, as long as the author and publisher are properly credited, which ensures maximum dissemination and a wider impact of our publications.

The book as a whole is distributed by MDPI under the terms and conditions of the Creative Commons license CC BY-NC-ND.

Contents

About the Editor vii

Michal Sedlačík
Advances in Elastomers
Reprinted from: *Materials* **2021**, *14*, 348, doi:10.3390/ma14020348 1

Miroslawa Prochon, Anna Marzec and Oleksandra Dzeikala
Hazardous Waste Management of Buffing Dust Collagen
Reprinted from: *Materials* **2020**, *13*, 1498, doi:10.3390/ma13071498 5

Md Mahbubur Rahman, Katja Oßwald, Katrin Reincke and Beate Langer
Influence of Bio-Based Plasticizers on the Properties of NBR Materials
Reprinted from: *Materials* **2020**, *13*, 2095, doi:10.3390/ma13092095 21

Marek Pöschl, Martin Vašina, Petr Zádrapa, Dagmar Měřínská and Milan Žaludek
Study of Carbon Black Types in SBR Rubber: Mechanical and Vibration Damping Properties
Reprinted from: *Materials* **2020**, *13*, 2394, doi:10.3390/ma13102394 37

**Sankar Raman Vaikuntam, Eshwaran Subramani Bhagavatheswaran, Fei Xiang,
Sven Wießner, Gert Heinrich, Amit Das and Klaus Werner Stöckelhuber**
Friction, Abrasion and Crack Growth Behavior of In-Situ and Ex-Situ Silica Filled
Rubber Composites
Reprinted from: *Materials* **2020**, *13*, 270, doi:10.3390/ma13020270 55

**Siti Aishah Abdul Aziz, Saiful Amri Mazlan, U Ubaidillah, Muhammad Kashfi Shabdin,
Nurul Azhani Yunus, Nur Azmah Nordin, Seung-Bok Choi and Rizuan Mohd Rosnan**
Enhancement of Viscoelastic and Electrical Properties of Magnetorheological Elastomers with
Nanosized Ni-Mg Cobalt-Ferrites as Fillers
Reprinted from: *Materials* **2019**, *12*, 3531, doi:10.3390/ma12213531 69

Sung Soon Kang, Kisuk Choi, Jae-Do Nam and Hyoung Jin Choi
Magnetorheological Elastomers: Fabrication, Characteristics, and Applications
Reprinted from: *Materials* **2020**, *13*, 4597, doi:10.3390/ma13204597 87

**Siti Khumaira Mohd Jamari, Nur Azmah Nordin, Ubaidillah, Siti Aishah Abdul Aziz,
Nurhazimah Nazmi and Saiful Amri Mazlan**
Systematic Review on the Effects, Roles and Methods of Magnetic Particle Coatings in
Magnetorheological Materials
Reprinted from: *Materials* **2020**, *13*, 5317, doi:10.3390/ma13235317 111

About the Editor

Michal Sedlačík received his Ph.D. degree from Tomas Bata University in Zlín (TBU) in 2012. From that same year, he has since been Lecturer at the Faculty of Technology at TBU. He successfully defended his habilitation work entitled "Novel Approaches to Design of Intelligent Fluids" in 2016. Currently, he is a senior researcher in the Nanomaterials and Advanced Technologies Group at Centre of Polymer Systems, TBU. He is a member of numerous international scientific societies such as American Chemical Society, The Society of Rheology, and The Society of Plastics Engineers. He shares authorship of 65 papers with h-index = 21. His current research interests include synthesis and properties of intelligent systems, elastomers, and electromagnetic shielding composites.

Editorial

Advances in Elastomers

Michal Sedlačík [1,2]

[1] Centre of Polymer Systems, Tomas Bata University in Zlín, Tr. T. Bati 5678, 760 01 Zlín, Czech Republic; msedlacik@utb.cz
[2] Department of Production Engineering, Faculty of Technology, Tomas Bata University in Zlín, Vavreckova 275, 760 01 Zlín, Czech Republic

Received: 7 January 2021; Accepted: 10 January 2021; Published: 12 January 2021

Elastomer materials are characteristic for their high elongation and (entropy) elasticity, which makes them indispensable for widespread applications in various engineering areas, medical applications or consumer goods. Their application-oriented development has to go hand in hand with material resources, economic aspects as well as environmental issues. Not only are elastomer matrix properties of high importance when optimizing the utility properties of the product; the fillers, plasticizers and other compounds are other property-determining factors that need to be considered, as these can positively affect the stimuli-responsive character, novel functionality, biological application, or fracture behavior, among other things, for gaining advances in elastomers [1–4].

This Special Issue, which consists of seven articles written by research experts in their field of interest, reports on the most recent research on rubber mixtures composition with the emphasis on the final utility properties of the product. Several novel and entrancing methods related to the friction and abrasion, waste management, bio-based composition, as well as two comprehensive reviews on the field-responsive elastomer materials, are introduced.

The environmental issue aspects in terms of hazardous waste management and bio-based resources are the subjects of the first two publications [5,6]. Prochon et al. used buffing dust collagen (BDC) originating from a waste product of the chromium tanning process in the leather industry as a modern filler in styrene butadiene rubber (SBR) [5]. The effect of this biodegradable filler on the physicochemical properties, biodegradation and thermo-oxidative aging of the SBR vulcanizates was investigated. The rod-like shape together with the scleroprotein additives presented on the surface of such a nanofiller led not only to more favorable dispersion within the elastomer, but also increased the cross-link density of the SBR vulcanizates, resulting in enhanced mechanical strength. Another advantage of BDC came from its antioxidant properties stabilizing (through the incorporation of chromium ions) the whole vulcanizate against thermo-oxidative aging, and its intense black color enabling it to be used also as a coloring additive. Finally, the release of compact chromium presented in the BDC filler from the vulcanizate is reduced by the formation of stable interfacial bonds between BDC and SBR.

Continuous material development is essentially important not only in the rubber industry. Rahman et al. proved the advantageous effect of bio-based plasticizers not only from the material resources viewpoint, but also as a processing aid positively affecting the complete process chain [6]. They investigated the suitability of two bio-based plasticizers, namely epoxidized esters of glycerol formal from soybean and canola oil, as sustainable alternatives with lower health risks compared to conventional plasticizers. Acrylonitrile-butadiene rubber with different ratios of monomers was used to observe the compatibility between the NBR and plasticizers for systems with different polarities. The positive effect of the bio-based plasticizers used in this study on the complete process chain was found in the NBR cure-accelerating effect, and the mechanical and thermal properties, which all were better in comparison with the conventional plasticizers-based systems investigated simultaneously.

Pöschl and co-workers followed the trend of eliminating the undesirable mechanical vibrations by vibro-insulating materials [7]. They investigated the ability to damp the mechanical vibration of SBR vulcanizates filled with four types of carbon black. The vibro-insulation tests comprised the forced oscillation method based on the transfer damping function. It was observed that the first resonant frequency decreased with an increase in the carbon black particle size as a higher transformation of input mechanical energy into heat under dynamic loading occurred. Durability experiments on the basis of harmonic loading revealed a shift of the first resonance frequency peak to lower excitation frequencies. The damping of mechanical vibrations had also been correlated with the excitation frequency of the vibration, the sample's thickness and inertial mass.

Rubber-like materials are frequently working under such conditions as rubbing, abrading, chunking, and tearing, which lead easily to mechanical failure. Vaikuntam et al. contributed with the friction, abrasion, and crack growth behavior of tire tread compounds composed of solution SBR filled with in-situ generated alkoxide silica or commercial precipitated silica [8]. It was demonstrated that a masterbatch containing in-situ silica could be used, with benefits for the industrial production of car tires or technical rubber articles in general, as its vulcanizates possessed a lower friction coefficient in comparison with classical precipitated silica-filled systems, which is promising for the low rolling resistance necessary, for example, for better fuel efficiency. Furthermore, if the samples having similar crosslink densities are compared, a much better resistance to crack growth was again observed for in-situ silica-based system.

For their characteristic properties, elastomer materials are used also in various advanced applications. Smart external field-responsive systems such as these can reversibly change their properties under an external stimulus, represented by an electric or magnetic field, pH, and light, among other things [9]. Aziz and co-workers led the world not only in the characterization of magnetorheological elastomers (MREs) reacting with changes in the systems' stiffness under the application of an external magnetic field, as is typical for such systems, but also included nanosized Ni–Mg cobalt ferrite particles into these MREs based on conventional carbonyl iron magnetic particles to make them suitable for use as actuators or flexible sensors [10]. Although the demanded electrical properties could be obtained with carbon-based particles, such as graphite or graphene, the necessary high concentration of the filler would negatively increase the stiffness of the system, thus the presented study could solve this problem. The developed MRE alters the magnetic, rheological and electrical resistance behavior appropriately, even at small concentrations (1.0 wt. %) of Ni–Mg nanoparticles.

The development of MREs is described in this Special Issue in two review papers complementing each other [11,12]. A more general review is given by Kang et al., in which the authors concentrate not only on the material basis used in MREs and covering a broad range of polymeric elastomeric materials, including, e.g., waste tire rubber as well, but also on the particles' distribution within the elastomeric matrix and its effect on the MR performance described in various experimental techniques [11]. Finally, in their review Jamari et al. concentrated on magnetic particles' coating as an important task necessary in order to obtain a system with improved sedimentation and oxidation stability, and appropriate tribology properties [12]. The chemical methods for particle coating, such as atom transfer radical polymerization, chemical oxidative polymerization or dispersion polymerization, are thoroughly described together with the effect of the chemical nature of the coating on the MR performance.

Funding: This work was supported by the Ministry of Education, Youth and Sports of the Czech Republic—project DKRVO (RP/CPS/2020/006).

Institutional Review Board Statement: Not applicable.

Informed Consent Statement: Not applicable.

Data Availability Statement: Data sharing not applicable.

Acknowledgments: First of all, I would like to express my deep gratitude to *Materials*, especially to its Editorial office, particularly to Andy Zhan, for continuous guide, care, and help during all steps of the preparation and production of this Special Issue entitled "Advances in Elastomers". I would also like to extend my gratitude to all contributing authors for their valuable manuscripts, as well as to all reviewers who have helped with valuable suggestions.

Conflicts of Interest: The author declare no conflict of interest.

References

1. Plachy, T.; Kratina, O.; Sedlacik, M. Porous magnetic materials based on EPDM rubber filled with carbonyl iron particles. *Compos. Struct.* **2018**, *192*, 126–130. [CrossRef]
2. Stocek, R.; Kipscholl, R.; Euchler, E.; Heindrich, G. Study of the Relationship between Fatigue Crack Growth and Dynamic Chip & Cut behavior of reinforced Rubber Materials. *KGK-Kautsch. Gummi Kunstst.* **2014**, *67*, 26–29.
3. Moucka, R.; Sedlacik, M.; Kutalkova, E. Magnetorheological Elastomers: Electric Properties versus Microstructure. *AIP Conf. Proc.* **2018**, *2022*. [CrossRef]
4. Cvek, M.; Moucka, R.; Sedlacik, M. Electromagnetic, magnetorheological and stability properties of polysiloxane elastomers based on silane-modified carbonyl iron particles with enhanced wettability. *Smart Mater. Struct.* **2017**, *26*, 105003. [CrossRef]
5. Prochon, M.; Marzec, A.; Dzeikala, O. Hazardous Waste Management of Buffing Dust Collagen. *Materials* **2020**, *13*, 1498. [CrossRef] [PubMed]
6. Rahman, M.M.; Osswald, K.; Reincke, K.; Langer, B. Influence of Bio-Based Plasticizers on the Properties of NBR Materials. *Materials* **2020**, *13*, 2095. [CrossRef] [PubMed]
7. Pöschl, M.; Vasina, M.; Zadrapa, P.; Merinska, D.; Zaludek, M. Study of Carbon Black Types in SBR Rubber: Mechanical and Vibration Damping Properties. *Materials* **2020**, *13*, 2394. [CrossRef] [PubMed]
8. Vaikuntam, S.R.; Bhagavatheswaran, E.S.; Xiang, F.; Wiessner, S.; Heindrich, G.; Das, A.; Stockelhuber, K.W. Friction, Abrasion and Crack Growth Behavior of In-Situ and Ex-Situ Silica Filled Rubber Composites. *Materials* **2020**, *13*, 270. [CrossRef] [PubMed]
9. Kutalkova, E.; Plachy, T.; Sedlacik, M. On the enhanced sedimentation stability and electrorheological performance of intelligent fluids based on sepiolite particles. *J. Mol. Liq.* **2020**, *309*, 113120. [CrossRef]
10. Aziz, S.A.A.; Mazlan, S.A.; Ubaidillah, U.; Shabdin, M.K.; Yunus, N.A.; Nordin, N.A.; Choi, S.B.; Rosnan, R.M. Enhancement of Viscoelastic and Electrical Properties of Magnetorheological Elastomers with Nanosized Ni-Mg Cobalt-Ferrites as Fillers. *Materials* **2019**, *12*, 3531. [CrossRef] [PubMed]
11. Kang, S.S.; Choi, K.; Nam, J.D.; Choi, H.J. Magnetorheological Elastomers: Fabrication, Characteristics, and Applications. *Materials* **2020**, *13*, 4597. [CrossRef] [PubMed]
12. Jamari, S.K.M.; Nordin, N.A.; Ubaidillah, U.; Aziz, S.A.A.; Nazmi, N.; Mazlan, S.A. Systematic Review on the effects, Roles and Methods of Magnetic Particle Coatings in Magnetorheological Materials. *Materials* **2020**, *13*, 5317. [CrossRef] [PubMed]

Publisher's Note: MDPI stays neutral with regard to jurisdictional claims in published maps and institutional affiliations.

© 2021 by the author. Licensee MDPI, Basel, Switzerland. This article is an open access article distributed under the terms and conditions of the Creative Commons Attribution (CC BY) license (http://creativecommons.org/licenses/by/4.0/).

Article

Hazardous Waste Management of Buffing Dust Collagen

Miroslawa Prochon *, Anna Marzec and Oleksandra Dzeikala

Institute of Polymer and Dye Technology, Faculty of Chemistry, Lodz University of Technology, Stefanowskiego 12/16, 90-924 Lodz, Poland; anna.marzec@p.lodz.pl (A.M.); dzeikala.sandra@gmail.com (O.D.)
* Correspondence: miroslawa.prochon@p.lodz.pl

Received: 24 February 2020; Accepted: 23 March 2020; Published: 25 March 2020

Abstract: Buffing Dust Collagen (BDC) is a hazardous waste product of chromium tanning bovine hides. The aim of this study was to investigate whether BDC has the desirable properties required of modern fillers. The microstructural properties of BDC were characterized by elemental analysis (N, Cr_2O_3) of dry residue and scanning electron microscopy (SEM). The BDC was applied (5 to 30 parts by weight) to styrene butadiene rubber (SBR), obtaining SBR-BDC composites. The physicochemical properties of the SBR-BDC composites were examined by Fourier transform infrared analysis, SEM, UV–Vis spectroscopy, swelling tests, mechanical tests, thermogravimetric analysis (TGA), and differential scanning calorimetry (DSC). The biodegradability of the SBR-BDC composites and their thermo-oxidative aging were also investigated. The filler contributed to increase the cross-link density in the elastomer structure, as evidenced by enhanced mechanical strength. The introduction of a filler into the elastomer structure resulted in an increase in the efficiency of polymer bonding, which was manifested by more favorable rheological and mechanical parameters. It also influenced the formation of stable interfacial bonds between the individual components in the polymer matrix, which in turn reduced the release of compact chromium in the BDC filler. This was shown by the absorption bands for polar groups in the infrared analysis and by imaging of the vulcanization process.

Keywords: buffing dust collagen (BDC); styrene–butadiene rubber (SBR); polymer composites; biodegradation

1. Introduction

The leather industry is one of the most polluting sectors of the economy [1]. Wastes from the leather industry represent 5% by weight of the raw materials [2]. The processing of leather, in particular tanning, also involves toxic chemicals, which can escape into the environment. The most widely used tanning method is chromium tanning (85–90% of world production), which is a type of chemical modification [3–5]. It uses chromium (III) salts in the form of chromium (III) sulfate, in combination with sodium sulfate. The tanning process imparts the skin with desirable properties, such as strength, thinness, and hydrothermal resistance [3,6], while also simplifying further processing. To ensure the durability and elasticity of the skin after tanning, it is subjected to a buffing process. A by-product of that process is buffing dust collagen (BDC). Buffing dust collagen is formed in the process of fat liquoring leather. According to the literature [6], chromium can attach to the active sites of collagen. Molecular modeling and IR analysis confirm that chromium can react with amino as well as carboxylate groups. Each ton of raw material generates ~0.6% buffing dust. If this type of waste goes to landfill, it may be hazardous for the environment, because the oxidation state of chromium salts changes from toxic III to VI, resulting in dangerous chromium salt products [4,7]. Another way to deal with such waste is incineration. Leather wastes have high calorific value (12–14 MJ/kg). However, sulfides are formed during their combustion, Volatile Organic Compounds (VOC) and greenhouse gases may be emitted to the atmosphere [5,8,9]. A final way of dealing with BDC waste is collagen extraction [2,10,11]. Due to the environmental impact of buffing dust and chrome shavings, tanning waste has become the object of intense research.

Studies have shown that BDC and chrome shavings collagen (CSC) can be successfully used as fillers in rubbers [12–17]. The dust form should simplify its dispersion in an elastomeric matrix [11]. Buffing dust can be introduced into natural latex rubbers and used as a filler in dust systems. Chromium strings have been applied in acrylonitrile butadiene (NBR) and styrene butadiene (SBR) rubber matrices [18,19]. Kowalska et al. [20] subjected leather waste from pork skins to alkaline reagents, which increased their polymer bonding efficiency and resulted in improved stabilization of interfacial interactions, thereby reducing the evolution of chromium. This led to improved mechanical parameters, which in turn increased the collagen added to PVC (polyvinyl chloride)-produced materials. Adding particles of buffing dust to poly(vinyl chloride) increased its Young modulus, the value of melt flow index (MFI), and susceptibility to biodegradation. Chrońska-Olszewska and Przepiórkowska [13] report that when applied in the form of a leather shavings/dust mixture to NBR and XNBR, the filler produced biodegradable collagen–elastomer materials with improved mechanical properties and hardness. The leather shavings/dust mixture was an active filler for NBR and XNBR.

Residues after tanning are not of particular value, and the Cr (III) salts may naturally turn into toxic Cr (VI) waste. Zhou et al. [21] used chromium-heated leather with active zirconium particles as a material for removing fluoride ions from groundwater. Research is also being carried out to transform tanning waste into carbon adsorbents at low temperatures, below 600 °C, e.g., using $ZnCl_2$ as the activating agent [22]. Leather waste has been used as a filling and stabilizing additive for bituminous and asphalt masses (Stone Matrix Asphalt (SMA)), improving their mechanical parameters, creep resistance, hardness, and humidity [23]. Ma et al. [24] developed a mesoporous material by high-temperature carbonization of chrome-tanned leather waste, which was then used for the electrodes in supercapacitors. The material was characterized by a high specific surface, low resistivity, and a high concentration of functional groups containing oxygen and nitrogen atoms. The admixture of leather waste with gravity substitutes for natural rubbers, acrylonitrile butadiene, or polyvinyl alcohol has led to interesting results [25,26]. The creation of hydrogen bonds is promoted, as well as chelation at interfaces, for example, between PVA and leather shavings, leading to greater compatibility of the tested centers. The elastomers of the tested rubbers also showed a significant increase in tear strength, due to the influence of skin particles.

The aim of the present study was to research the effect of applying buffing dust as a filler to styrene–butadiene rubber. There have been no previous reports of using BDC sanding dust from chrome tanning processes as a filler for SBR rubber in combination with a conventional seeding unit. Before being introduced into the elastomer matrix, the BDC was characterized by FTIR, SEM analysis, elemental analysis (including determination of chromium content), dynamic light scattering (DLS), and the DBP test. The crushed dust collagen reacted with the elastomer matrix and other components in the mixtures, as was confirmed by FTIR and mechanical studies.

2. Materials and Methods

2.1. Materials

2.1.1. Rubber and Other Ingredients

The continuous phase was Styrene–Butadiene Rubber (SBR), KER 1500, from Bayer AG Company, Leverkusen, Germany. The other components were zinc oxide pure (ZnO) from LANXESS Deutschland GmbH, Augsburg, Germany; sulfur pure (S_8) (density 2.07 g/cm^3) from Siarkopol Tarnobrzeg Sp. z o.o., Tarnobrzeg, Poland; technical stearin, from Torimex Chemicals Ltd. Sp. z o.o., Konstantynów Łódzki, Poland; MBTS pure from Accelerator Bayer AG Company, Leverkusen, Germany; and toluene from Chempur Company, Piekary Śląskie, Poland.

2.1.2. Buffing Dust Collagen (BDC)

Buffing dust was sourced as a waste product from various batches of cattle skins from Kalisz Tabbery, Kalskór S.A., Kalisz, Poland. The black dust (i.e., basic chromium sulfate $Cr(OH)SO_4$, chromium salts, vegetable tannins, etc.) was generated as a result of grinding tanned leather in the final stage of leather production, after dyeing and greasing, in a chromium system. The BDC was mixed, sieved through 2 mm sieves on a vibrating screen (AS200 Control, Retsch GmbH, Haan, Germany), and then conditioned at 50 °C for 5 h in a Binder thermal chamber (Binder GmbH, Tuttlingen, Germany). The content of chromium (III) as Cr_2O_3 varied from 4.25% to 4.48%, in accordance with the PN-EN standard ISO 4684: 2006 (U) [13,14]. The BDC fiber had a diameter of 0.2 mm.

2.2. Preparation of Composites

After the preparation step (see Section 2.1.2), the BDC fillers were applied to the styrene butadiene rubber at different concentrations. The rubber mixtures were prepared using a mixing mill (Bridge type milling machine, London, UK) with a roll temperature of 27–37 °C and friction of 1.1. The parameters of the rolling mill are as follows; roller length: L = 450 mm; roll diameter: D = 200 mm; rotational speed of the front roller: Vp = 20 rpm; width of the gap between the rollers: 1.5–3 mm. The mixtures were prepared for 6 min, then packed in foil and stored at 2–6 °C. The compositions of the rubber composites are given in Table 1. The tests were carried out at room temperature under normal pressure.

Table 1. Composition of the Styrene–Butadiene Rubber (SBR) composites.

Symbol	SBR	SBR5	SBR10	SBR20	SBR30
SBR (phr)	100	100	100	100	100
BDC (phr)	0	5	10	20	30
ZnO (phr)	5	5	5	5	5
Sulphur (phr)	2.5	2.5	2.5	2.5	2.5
MBTS (phr)	1.5	1.5	1.5	1.5	1.5
Stearic acid (phr)	2	2	2	2	2

Vulcanizates were prepared in hydraulic presses. Samples in the form of 100 × 50 mm films were prepared by pressing under 150 MPa at 150 °C.

3. Research Techniques

The curing kinetics of the SBR compounds were studied in a moving die rheometer (MonTech MDR 300, Buchen, Germany) at 150 °C, according to the ISO 3417 standard. The rubber compounds were vulcanized according to the optimal curing time (τ_{90}) in a standard electrically heated hydraulic press at a temperature of 150 °C, with a pressure of 15 MPa. The mechanical properties of the prepared composites were tested using a Zwick universal testing machine, model 1435 (according to PN-ISO 37:1998). The tensile strength (TS_b) and percentage elongation at break (E_b) were determined. Hardness testing (H, °Sh) was carried out using a Shore electronic hardness tester, type A, with a force of 12.5 N, according to standard PN-80C-04238 (Zwick/Roell, Herefordshire, UK).

The polymer–solvent interaction parameter (0.378 for SBR rubber in toluene solvent) was determined based on the equilibrium swelling method (according to PN-ISO 1817:2001/ap1:2002). The cross-link density (ν, 10^4 mol/dm^3) was calculated as the volume fraction of rubber in the swollen material, and $V_S = 106.3$ mol/cm^3 for the molar volume of solvent (toluene) [13–17,27].

Cross-linking density, ν, was calculated on the basis of Flory–Rehner's Equation (1):

$$\nu = \frac{\left[\ln(1 - V_r) + V_r + \mu V_r^3\right]}{V_O\left(V_r^{\frac{1}{3}} - \frac{V_r}{2}\right)} \quad (1)$$

where μ is the Huggins parameter for the uncross-linked polymer–solvent system and V_r is the molar volume of the swelling solvent.

The thermo-oxidative experiments were performed in a convection oven and a thermal chamber with air circulation (Binder GmbH, Tuttlingen, Germany). An unstressed sample was exposed to the action of circulating air at 70 °C for 168 h. To study the deformation energy of the vulcanizates as a result of biological aging, the aging factor (S) was calculated according to Equation (2):

$$S = \frac{TS_{b1} \times E_{b1}}{TS_{b2} \times E_{b2}} \quad (2)$$

where $TS_{b1} \times E_{b1}$ is the tensile strength (MPa) and elongation at break (%) after thermal-oxidative aging or soil test and $TS_{b2} \times E_{b2}$ is the tensile strength (MPa) and elongation at break (%) before thermal-oxidative aging or the Soil Test.

A biodecomposition test was performed in soil with paddle-shaped samples with dimensions of 7.5 cm by 1.25 cm, and sampling of 0.4 cm. The samples were placed in an active universal soil (10 cm depth) and incubated at a temperature of 30 °C with 80% RH for 90 days in a climatic chamber (HPP 108 Memmert GmbH, Schwabach, Germany). Tests were carried out according to PN-EN ISO 846. The soil test was analyzed following the method described by Tadeusiak et al. [15]. Surface topography of the composites was conducted after the soil tests, using photos taken with a Canon CanoScan 4400F device. The morphology of the BDC powder and SBR composites were analyzed using a scanning electron microscope (SEM), Zeiss Ultra Plus (Bruker). Prior to the analysis, the samples were coated with a carbon target using a Cressington 208 HR system [13,16]. A Nicolet 6700 FT-IR spectroscope (Thermo Scientific, Waltham, MA, USA) with Fourier transformation and ATR snap was used to determine the characteristics of the composites. Analysis performed in the range of 4000 to 400 cm^{-1} [13]. Differential scanning calorimetry (DSC) was performed using a DSC1 analyzer (Mettler Toledo, Netzsch, Switzerland) at a heating rate of 10 °C/min. The SBR samples were heated from −150 °C to 350 °C under a nitrogen atmosphere. Thermogravimetric analysis (TGA) was performed using a TGA/DSC1 analyzer (Mettler Toledo, Netzsch, Switzerland). The heating rate was 10 °C/min under a nitrogen atmosphere, across a temperature range of 25 to 900 °C. DSC was analyzed as described by Prochoń et al. [28]. The changes in color of the samples after the thermo-oxidative aging process and soil test were studied using a UV–Vis CM-3600d spectrophotometer (Konica Minolta, London, UK). The difference in color was expressed as the color change parameter dE × Lab, where L is the level of lightness or darkness, a is the relationship between redness and greenness, and b is the relationship between blueness and yellowness [13].

Elemental analyses of the carbon, hydrogen, and nitrogen elements in the BDC powder were carried out using a Vario EL III analyzer equipped with special adsorption columns and a thermal conductivity detector (TCD). The absorption of dibuthylphtalate (DBP) by the BDC powder was measured using an Absorptometer C (Brabender mixer, Brabender GmbH & Co. KG, Duisburg, Germany). The sizes of the BDC particles were determined in water (filler concentration of 0.01 g/250 mL) using the dynamic light scattering (DLS) method on a Zetasizer Nano (Malvern Instruments, Malvern, Great Britain) analyzer [14].

4. Results and Discussion

4.1. Characterization of BDC Powder

The valence vibrations of bands derived from methyl and methylene groups in the tested BDC dust are visible in the wave number range of 3420 to 3479 cm^{-1} (Figure 1). Maxima of the bands appeared in the dust spectrum at a lower wave number intensity. Bands at 1750 cm^{-1} (–COO$^-$) indicate the presence of fatty substances [13,14,28].

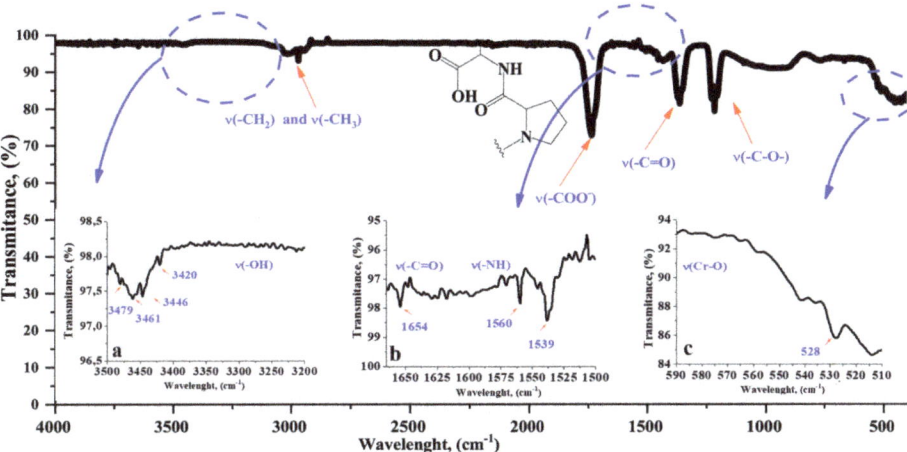

Figure 1. IR spectra of buffing dusts of collagen with ranges: (**a**) 3500–3200 cm^{-1}; (**b**) 1650–1500 cm^{-1}; (**c**) 590–510 cm^{-1}.

In the area from 3200 to 3500 cm^{-1} there is a wide band corresponding to the valence vibrations of hydroxyl (–OH) side chains and end groups, with higher intensity in the dust spectrum. In the range of 1650 to 1500 cm^{-1}, a band of deformation vibrations of the first-order amide appears for (C=O) and second-order amide (NH). The effect of chromium on interactions with other components of the BDC powder, and thus other interactions, was visible through shifts in the absorption bands, reducing their intensity, etc. The characteristic absorption band at 1654 cm^{-1} may be attributed to the possible mechanism of interaction of Cr with the protein-like system –Cr–OOC– (Figure 1b) [6]. The band of COOH stretching vibrations derived from amide I is shifted from 1660 cm^{-1}. There is also a characteristic wide absorption band for chromate samples at about 1000 cm^{-1}, which formed as a possible result of the interaction of chromium with a carboxyl group. The presence of Cr–O–Cr bonds is indicated by the bands between 510 and 650 cm^{-1} [6,13].

The shape, particle size, and specific surface of the filler are known to have a decisive impact on the strength of rubber–filler joints. Dust morphology was assessed based on photos taken using SEM, as shown in Figure 2. Dust agglomerates are visible as primary particles with a regular structure: elongated, insulated fibers with a wide size distribution from several hundred nanometers to several micrometers.

One collagen macrofibrillary fiber is connected to several or even several dozen individual helical microfibers (Figure 2a), with diameters of ~3–4 micrometers and longitudinal segmentation.

The particle size distribution of the tested BDC was measured by dynamic light scattering (DLS) in an aqueous solution. The particle size distribution, which is a compilation of measuring the length and diameter of particles oriented in the laser light field, was in the range from 469 to 295 nm. The isoelectric Point (IEP) was at pH 5.9. There was an appropriately small area of 9 m^2/g. However, based on the elemental analysis of BDC, the nitrogen conversion to protein substance, which determines the nitrogen content in the collagen, was 7.92%. The Cr converted to Cr_2O_3 was at 4.48%. Dry matter was 89.49%, and ash was ~7% [13,14].

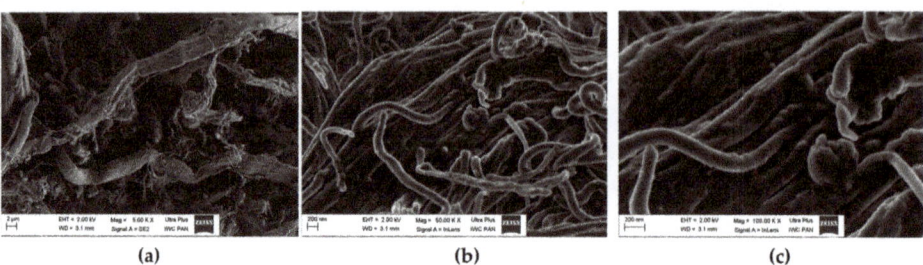

Figure 2. SEM images of buffing dust collagen (BDC): (**a**) 5000× magnification, (**b**) 50,000× magnification, and (**c**) 100,000× magnification.

Oil number is one of the important factors that measure the structure and surface of fillers. From a morphological point of view, fillers have the ability to form aggregates or agglomerates. To prevent the influence of physico-chemical interactions, fillers are often subjected to modifications aimed at changing their structural or surface characteristics. The structure of the BDC had been changed under the influence of the chemical modification processes during tanning.

As can be seen in the Annex (see Supplementary Materials Figure S1), after the addition of dibutyl phthalate (DEP), the BDC molecules begin to approach each other and form agglomerates. There is increasing resistance to mixing, due to the higher torque. At the moment of maximum saturation with PBT, the process ends, obtaining maximum torque. The maximum torque was 248.4 mNm. The filling volume efficiency was determined using the Medalia model, according to Equation (3):

$$\varphi_{eff} = 0.5\varnothing \left[1 + \left(1 + \frac{0.0213(\text{DBP})}{1.46}\right)\right] \quad (3)$$

where φ_{eff} is the actual volume of the filler. The efficiency of the BDC was 4.33 mL/g. The moisture content of the BDC particles was in the range of 10^2 to 10^4 nm. This important parameter classifies the BDC filler in the group of semi-reinforcing fillers. The filler shows a high degree of orderliness and a tendency to form aggregates.

4.2. Characterization of SBR/BDC Composites

4.2.1. Rheometric Properties

The BDC filler clearly affected the curing characteristics of the SBR mixtures. The kinetic parameters of the BDC-based compounds differed significantly from those of the unfilled compound, as shown in Table 2. The incorporation of BDC into the SBR compound resulted in higher viscosity and stiffness, as reflected in the torque values (LH, ΔL), which are much higher than those of the SBR samples. The incorporation of the biofiller resulted in higher cross-link density, as the ΔL parameter is an indirect measurement of the degree of elastomer cross-linking. The unfilled sample exhibited shorter time of vulcanization and scorch time than the filled composites. The scorch and vulcanization times increased with increasing concentrations of BDC, showing the highest values for 30 phr concentration (τ_{02} = 3.5 and τ_{90} = 38.5). It can be concluded that the sulfur cross-linking system was partially adsorbed onto the outer surface of the BDC filler, resulting in a slower curing process [12,13].

Table 2. Influence of BDC on the rheometric properties of SBR compounds.

Symbol	L_L (dNm)	ΔL (dNm)	τ_{02} (min)	τ_{90} (min)
SBR	11.0 ± 0.3	39.9 ± 1.1	1.2 ± 0.1	24.5 ± 1.3
SBR5	13.9 ± 0.1	54.8 ± 1.9	2.9 ± 0.1	30.0 ± 1.4
SBR10	12.0 ± 0.2	61.6 ± 2.2	2.0 ± 0.1	33.0 ± 1.1
SBR20	13.0 ± 0.1	51.7 ± 1.9	2.9 ± 0.2	37.0 ± 1.0
SBR30	14.0 ± 0.1	47.5 ± 1.4	3.5 ± 0.2	38.5 ± 1.2

L_L—minimum torque moment (dNm); ΔL—the decrease of torque moment (dNm) ($\Delta L = L_{HR} - L_L$); τ_{02}—scorch time (min); τ_{90}—time of vulcanization (min).

4.2.2. Cross-Linking Density

Figure 3 shows the effect of BDC on the cross-linking density of the SBR composites. Increasing the concentration of the protein filler caused a gradual increase in the degree of cross-link density in the SBR vulcanizates. These results are in agreement with the previous rheometer measurements.

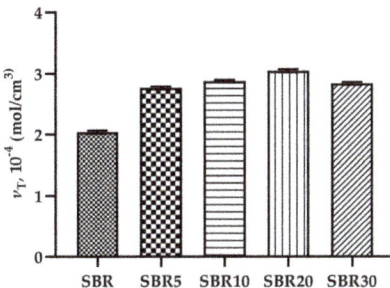

Figure 3. Influence of BDC on the cross-linking density of SBR vulcanizates.

The higher degree of cross-link density in the SBR/BDC composites can be explained by the reactivity of groups originating in the BDC filler and the elastomeric matrix. The most probable interactions occur between the styryl group in rubber and the polar fragments of the collagen dust, as shown in Figure 4. The interactions between the BDC dust and the elastomer matrix were also confirmed by infrared analysis (Figures 5 and 6).

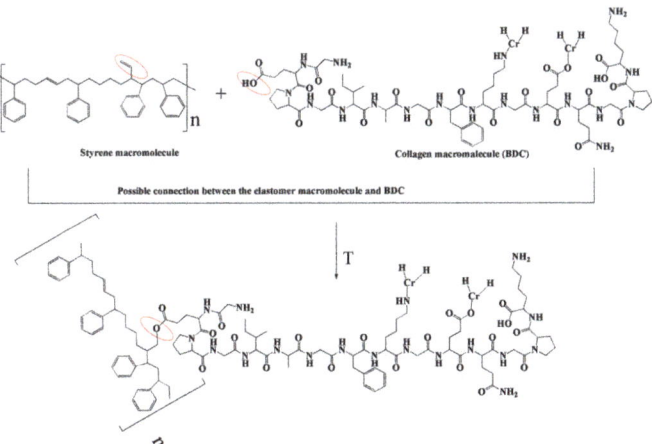

Figure 4. Possible mechanism of interaction between the elastomer macromolecule and BDC structure.

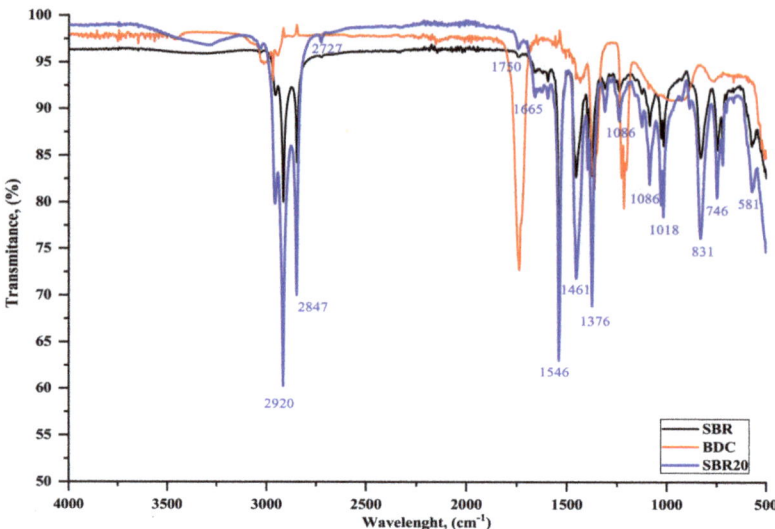

Figure 5. Fourier transform infrared spectroscopy (FTIR) spectrum of samples containing 0 and 20 phr BDC in SBR composites relative to the BDC spectrum.

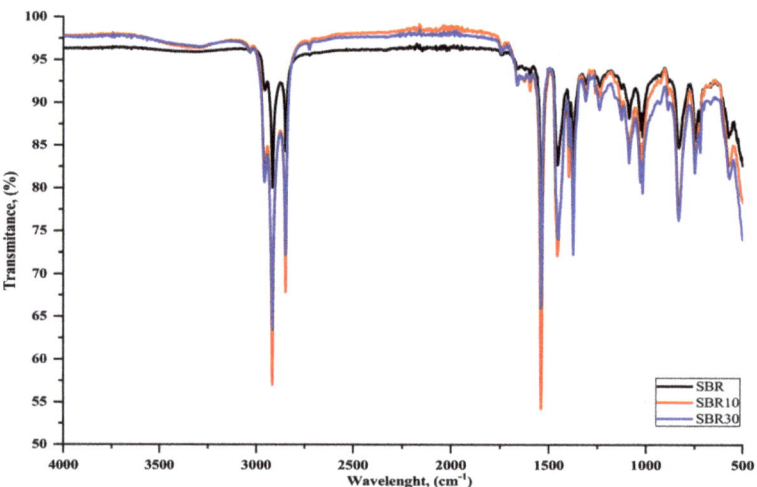

Figure 6. Comparison of the FTIR spectra for the standard sample (SBR), 10, and 30 phr BDC in SBR composites.

4.2.3. FTIR Analysis

In the next part of the study, the composites were subjected to FTIR analysis. Figures 5 and 6 show the sample spectra of SBR composites containing various amounts of grinding dust, as well as the reference spectrum. Based on the BDC spectra of both SBR composites, a small band can be seen in the 3400 to 3250 cm^{-1} range (slightly larger for SBR20), which is associated with vibrations of the -OH group (Figure 5). In the range of 3100 to 3000 cm^{-1}, a peak is visible from the stretching vibrations of the C–H bonds of the aromatic ring that form part of the SBR vulcanizate side groups [29]. In the 2950 to 2800 cm^{-1} range, intense peaks are visible from the vibrations of the C–H groups present in the material, both from the styrene aromatic ring and from the protein chain. The peak in the range of 1700 to 1630 cm^{-1} is caused by the presence of C=C diene groups. The visible change and shifts in absorption bands at 1750 to 1730 cm^{-1} can also be associated with the chromium complex and the free carboxyl group of oligomers that is generated by the fragmentation of peptide chains. The bands in the 1665 to 1550 cm^{-1} range are derived from the vibrations of the C=O peptide bond groups [13,14]. The deviation and significant decrease in the intensity of the absorption band at 1665 cm^{-1}, as well as its shift to 1630 cm^{-1} in the SBR or SBR20 spectra relative to the spectrum for BDC, is caused by chromium, which can be coordinated with active collagen centers (amides I and II), thus forming a chromium complex [6]. Together with the shift, this significant reduction in the band (1730 cm^{-1}) for the BDC spectrum compared to that for SBR20 supports our supposition that the probable mechanism of interaction between the elastomer macromolecule and BDC structure is as shown in Figure 4 [6].

The intensity of the BDC bands also changes depending on the amount of dust. In comparison with the spectrum for BDC alone (Figures 1 and 6), the intense band at 1665 cm^{-1} derived from primary and secondary amides –(H$_2$N)CO; –R(NH)CO significantly reduces its intensity, which may indicate the possibility of interactions with the C-C groups in the SBR aromatic rubber ring. The absorption band in the 1600 to 1550 cm^{-1} range is also caused by the presence of C–C bonds in the aromatic ring. The intense bands at 1546 cm^{-1} are derived from the stretching vibrations of the –CH$_2$ and –CH$_3$ groups. The small band around 1300 cm^{-1} is from H–O–H groups. It is associated with the presence of the small amount of water in zinc oxide. The band between 1018 and 1080 cm^{-1} is related to the absorption of stretching vibrations originating from C–C moieties in the aromatic ring, which are much more intense when BDC is applied to the SBR structure. This may indicate some interaction between the filler and the elastomeric matrix. The clear peak around 831 cm^{-1} is responsible for the presence of disubstituted alkenes. Intense vibrations are visible at 746 cm^{-1}, which may be due to vibrations by hydroxylysine N–H groups present in the BDC or at 581 cm^{-1} from sulfur groups S–H or S–S [13,30].

4.2.4. Mechanical and Hardness Tests

The effects of the BDC on the mechanical properties of the SBR vulcanizates are presented in Table 3. The incorporation of the protein filler enhanced the tensile strength (TS) of the SBR composites in almost all cases, except for sample SBR5 containing the lowest amount of BDC (5 phr). The application of BDC significantly increased the stiffness of the composites, as reflected in the higher stress at 300% elongation (SE300). This parameter is related to the cross-link density of the composites and confirms the results from the previous swelling measurements and rheometric studies. The highest values for tensile strength and SE300 modulus were noted when 10 phr of BDC was added. These values then decreased upon further addition of the protein filler. This may be related to the greater tendency of hydrophilic BDC filler to agglomerate at higher concentrations in the SBR matrix. After the addition of BDC, the elongation at break was significantly reduced in all of the BDC-based composites, most likely due to the increase in crosslink density. These observations are in agreement with rheometric studies and swelling experiments.

Table 3. Influence of BDC filler on the mechanical properties of SBR composites.

Parameter	SBR	SBR5	SBR10	SBR20	SBR30
TS (MPa)	3.34 ± 0.1	3.73 ± 0.2	4.78 ± 0.3	4.91 ± 0.3	3.94 ± 0.2
SE300 (%)	2.15 ± 0.5	2.41 ± 0.1	2.60 ± 0.2	3.07 ± 0.6	3.70 ± 0.1
E$_b$ (%)	541 ± 4.2	228 ± 6.1	254 ± 5.5	322 ± 7.8	267 ± 4.9
H (°Sh)	44.08 ± 0.9	48.22 ± 1.2	49.20 ± 1.1	50.85 ± 0.8	53.88 ± 0.7

The incorporation of BDC in SBR improved the hardness of the prepared composites in comparison to the reference sample (Table 3). The hardness increased with increasing filler content. The most pronounced increase in hardness was observed in the vulcanizate containing 30 phr of BDC filler. The higher hardness values of the SBR/BDC samples can be explained by the presence of the BDC filler and the increased cross-link density of the composites. Compared with other studies [12–14] where buffing dust BDC was introduced into, e.g., a matrix of rubbers with special applications, namely, acrylonitrile butadiene rubber (NBR) and carboxylated acrylonitrile butadiene rubber (XNBR), there was an increase in the tensile strength and elongation at break. This could be related to the different ways in which the buffing dust was applied to the polymer matrices. In the previous studies, BDC was introduced by direct mixing with a cross-linking process activator, zinc oxide (ZnO).

4.2.5. SEM Analysis

Filler dispersion is an important factor that affects the properties of rubber composites. Therefore, the next stage of the study investigated the distribution of BDC in the SBR matrix. Figure 7a–e presents SEM micrographs of cross sections of the SBR/BDC composites at different magnifications. The micrographs of SBR10 and SBR30 composites at 1.00× magnification show a rather homogeneous distribution of components in the elastomeric matrix. The reference sample showed fine, suspended particles of curing components, which are also visible in the photos of composites with the addition of BDC. The magnifications at 50.00 and 100.00 for BDC-based samples show irregular fragments of suspended particles, with diameters of approximately 100 or 200 nm. It appears that the agglomerates in the sample containing 10 phr of BDC were smaller in size than those in the SBR30 composite.

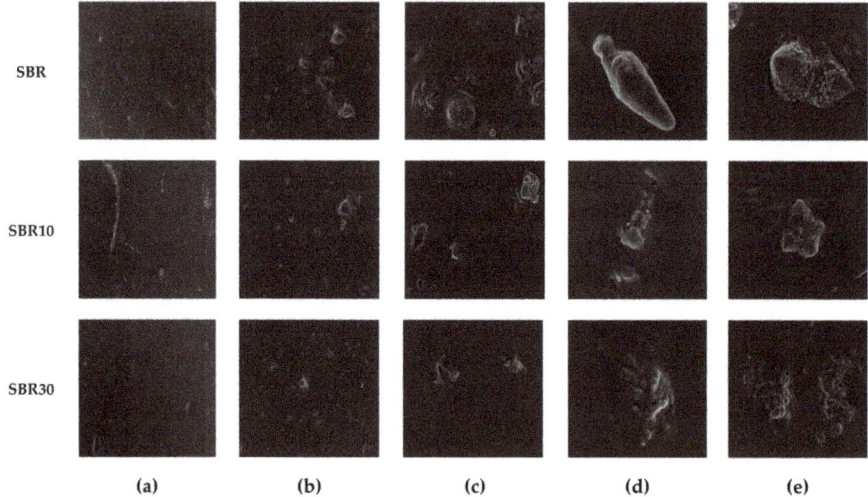

Figure 7. SEM images of SBR, SBR10, and SBR30 composites at different magnifications: (**a**) 1.00×, (**b**) 10.00×, (**c**) 25.00×, (**d**) 50.00×, and (**e**) 100.00×.

4.2.6. DSC & TGA analysis

The results of the thermogravimetric analysis (TGA) and differential scanning calorimetry (DSC) of selected SBR and SBR10 composites are shown in Table 4, as well as in Figures 8 and 9.

Table 4. Thermal characteristics of SBR and SBR10.

Symbol	$T_{5\%}$ (°C)	$T_{peak\,(DTG)}$ (°C)	Δm_{total} (%)	T_g (°C)
SBR	305	478.77	95.32	−42.08
SBR10	299	479.64	90.34	−45.45

$T_{5\%}$—decomposition temperature at 5% mass loss, $T_{p(DTG)}$—temperature of maximum conversion rate on the DTG curve, Δm_{total}—total mass loss during thermal decomposition, T_g—glass transition temperature (standard deviations: T_5, $T_{p\,(DTG)} \pm 2$ °C; $\Delta m_{total} \pm 0.6\%$; $T_g \pm 2$°C)

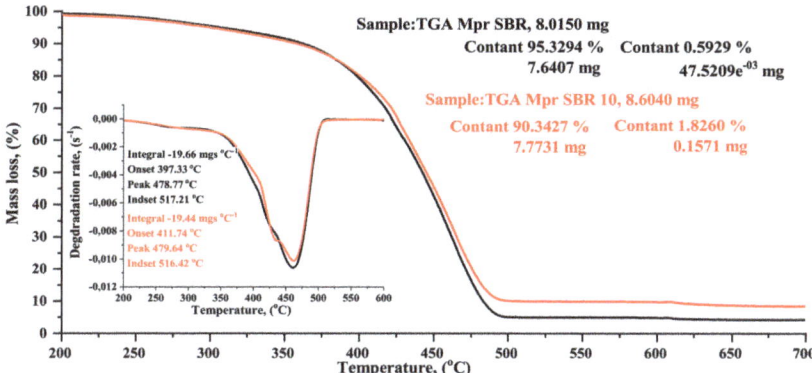

Figure 8. Thermogravimetric analysis (TGA) curves of SBR composites: reference sample SBR (black curve) and SBR10 sample (red curve).

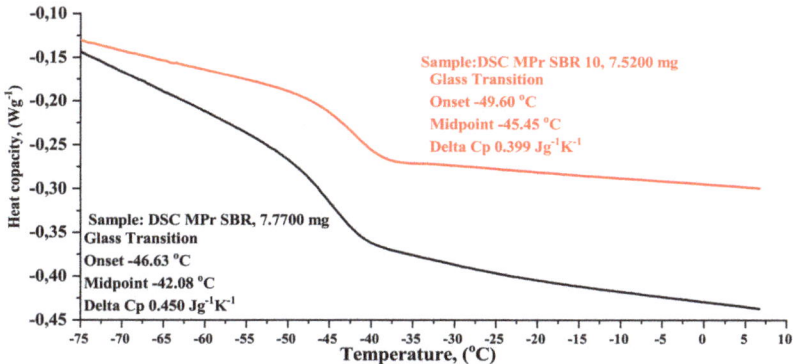

Figure 9. Comparative analysis of DSC results: SBR—unfilled (black curie) and SBR10 (10 phr of BDC—red curve).

The introduction of the BDC filler to SBR did not significantly affect the temperature of thermal decomposition. In thermogravimetric analysis, a 5% weight loss was observed for the SBR10 composite at a slightly lower temperature than the reference sample. As can be seen from the standard deviation, the maximum decomposition temperatures of all the composites are similar. The total weight loss of the tested elastomers during decomposition was 95% for the reference sample and 90–91% for the composites filled with BDC.

The glass transition temperature for pure styrene–butadiene rubber is reported in the literature data as 48/65 °C [31]. The DSC revealed that the glass transition temperature T_g begins for the SBR composite at onset, i.e., at −46.6 °C, whereas for SBR10, T_g begins at −49.60 °C (Table 4, Figure 9). The glass transition temperature is shifted relative to the reference composition (SBR, −42.08 °C) towards lower values for the composition containing BDC dust (−45.45 °C). This shift is clear in the case of the sample with 10 parts wt. BDC. It suggests that BDC filler acts as a plasticizer in SBR compounds. The heat capacity of the SBR composite was 0.450 Jg^-1K^-1, whereas for SBR10, heat capacity decreased to 0.399 Jg^-1K^-1.

4.2.7. Swelling Balance in Water, Color Research, Soil Test, and Thermo-Oxidative Aging Process

There is considerable interest in accelerated aging processes and the behavior of polymer materials in contact with various factors. The dust-based elastomer composites were therefore subjected to water infraction into the material structure (Q_w, -), soil tests, color change, and accelerated thermo-oxidative aging (ΔT, -). Figure 10 presents the results of equilibrium swelling in water. The addition of protein filler caused swelling extension relative to the unfilled sample. Due to the hydrophilic character of protein, higher water absorption was observed. The absorption of water facilitated the penetration of microorganisms, and thus the biodegradation process was more rapid.

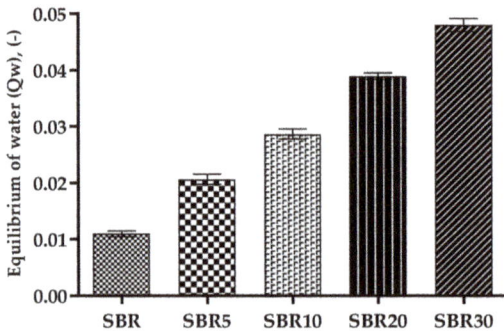

Figure 10. Equilibrium of swelling in water results from SBR vulcanizates.

The next test was to determine the surface color change (dE × ab, -) of the vulcanizates exposed to elevated temperature and biological properties (soil tests). Table 5 shows photographs taken before and after biodegradation. As can be seen, topographic changes occurred in the composites containing the bionic biofiller. There were discolorations in the material where micro- and macro-cracks formed on the surface. The change in the appearance was influenced by the action of the biological agents found in soil, which, by breaking down the protein filler, contributed to the deterioration of the properties of the tested vulcanizates.

Table 5. Topography photos of surface changes of SBR composites before and after biodegradation tests.

Composite Symbol	SBR	SBR5	SBR10	SBR20	SBR30
BEFORE BIODEGRADATION					
dL × ab	3.84	0.99	0.20	0.26	0.46
AFTER BIODEGRADATION					
dL × ab	7.35	2.63	1.34	4.52	3.18

Factor dL × ab was found to change the color of the sample after degradation relative to the sample before degradation. The highest values for dL × ab were found for the native composite, SBR, which proves that the other samples were protected against thermo-oxidative aging, especially in the case of the 10 or 20 phr biopolymer. Higher values for this indicator are associated with greater color deviation in the case of biodegradation. The action of soil organisms promotes the degradation of the filler, which leads to a loss of consistency in the material structure. This indicates much greater susceptibility to biodegradation in the case of vulcanizates containing the protein filler. The changes in the color of the composites were affected by the aging conditions, i.e., temperature, humidity, as well as microorganisms contained in soil in soil tests.

It was found that addition of protein filler resulted in acceleration of biodegradation in almost all of the samples. The mechanical deterioration of composites with BDC dust to the soli test can also be seen on the basis of the aging coefficient (ΔT, -), as the combination of tensile strength and elasticity before and after aging (Table 6).

Table 6. Mechanical properties after biodegradation.

Effects of Aging and Biological on the Properties of SBR Composites					
Composite symbol	SBR	SBR5	SBR10	SBR20	SBR30
Aging coefficient ΔT(-)					
Thermal aging	0.93 ± 0.04	1.42 ± 0.02	0.84 ± 0.03	1.14 ± 0.01	0.92 ± 0.03
Soil test	0.94 ± 0.01	0.83 ± 0.01	0.67 ± 0.02	0.56 ± 0.01	0.51 ± 0.02
Hardness of obtained composites H (°Sh)					
Thermal aging	45.17 ± 0.12	51.95 ± 0.08	53.33 ± 0.14	54.25 ± 0.23	58.15 ± 0.18
Soil test	45.00 ± 0.20	46.90 ± 0.13	48.07 ± 0.21	49.27 ± 0.10	53.03 ± 0.09

On the basis of soil tests, it appears that biodegradation caused a decrease in tensile strength (Tables 5 and 6). This decrease is probably due to filler degradation, which causes a reduction in intermolecular bonding [32]. It was found that the addition of the protein filler resulted in accelerated biodegradation in almost all the samples. The hardness of samples was greater after the soil tests than before the tests. The greater hardness was a consequence of the biodegradation process. Increased hardness was noticed as a result of thermo-oxidative aging relative to the non-aged samples. The noticeable increase in hardness may be associated with greater cross-link density, secondary cross-linking, or the evaporation of water derived from the hydrophilic filler. Increasing the filler load resulted in higher hardness values.

5. Conclusions

In this study, collagen grinding dust was applied as a biodegradable filler in the SBR polymer, to study its effects on the mechanical properties, biodegradation, and thermal stability of the SBR vulcanizates. Due to its particle size of less than 1 micrometer, BDC has potential to be used in SBR rubber as nanofillers. The elongated rod-like shape promotes more favorable dispersion of the filler in the elastomer medium, as shown by SEM photos. The presence of a scleroprotein additive results in a higher degree of cross-linking in the case of vulcanizates, which in turn increases the stiffness and hardness of the composites. Interactions may occur between the components of the elastomer matrix, as evidenced by the increased intensity of IR absorption bands in the range of 1000 to 1750 cm^{-1}. A greater weight share of filler results in shift of glass transition temperatures towards lower values, and improved hardness. As a result of the biological or thermal aging with oxygen, greater water infiltration was observed in the structures of the material produced on the basis of SBR30 rubber. Surface color stabilization was also visible, e.g., under the influence of ultraviolet radiation. The introduction of BDC dust had a stabilizing effect on thermo-oxidative aging processes, due to the antioxidant properties of collagen dust itself. The obtained studies demonstrated that sifted and thermostabilized grinding dust can play a role as a biodegradable filler in polymer materials and, due to intense black color, can act as coloring additive. As expected, the thermal stability of the composites increased, most likely due to the incorporation of chromium ions in the polymer network. The results of this study expand the possibilities for managing hazardous tanning wastes in the form of dust from chrome tanned leathers, while reducing their environmental impact.

Supplementary Materials: The following are available online at http://www.mdpi.com/1996-1944/13/7/1498/s1, Table S1: BDC Dust characteristic characteristics; Figure S1: BDC Oil number (the sample weight was 4.3 g at 125 rpm and the dosing frequency was 4.0 mL/mL).

Author Contributions: Conceptualization, investigation, writing—original draft preparation, formal analysis, M.P.; software, methodology, resources, M.P. and O.D.; visualization, project administration O.D.; supervision, validation A.M. All authors have read and agreed to the published version of the manuscript.

Funding: This research received no external funding.

Conflicts of Interest: The authors declare no conflicts of interest.

References

1. Ravindran, B.; Wong, J.W.; Selvam, A.; Thirunavukarasu, K.; Sekaran, G. Microbial biodegradation of proteinaceous tannery solid waste and production of a novel value added product–Metalloprotease. *Bioresour. Technol.* **2016**, *217*, 150–156. [CrossRef]
2. Masilamani, D.; Madhan, B.; Saravanan, G.; Palanivel, S.; Narayan, B. Extraction of collagen from raw trimming wastes of tannery: A waste to wealth approach. *J. Clean. Prod.* **2016**, *113*, 338–344. [CrossRef]
3. Duperriez, F.; Poncet, T.; Lety, R.; Kulińska, I.; Kosińska, K.; Mikulska, M.; Prygiel, H.; Sadowski, T. *Przewodnik Wprowadzania Systemu Zarządzania środowiskowego w Garbarniach*; AFNOR: Paris, France, 2010; pp. 11–13.
4. Bacardit, A.; Baquero, G.; Sorolla, S.; Ollé, L. Evaluation of a new sustainable continuous system for processing bovine leather. *J. Clean. Prod.* **2015**, *101*, 197–204. [CrossRef]
5. Bacardit, A.; van der Burgh, S.; Armengol, J.; Ollé, L. Evaluation of a new environment friendly tanning process. *J. Clean. Prod.* **2014**, *65*, 568–573. [CrossRef]
6. Nashy, E.H.A.; Osman, O.; Mahmoud, A.A.; Ibrahim, M. Molecular spectroscopic study for suggested mechanism of chrome tanned leather. *Spectrochim. Acta Part A Mol. Biomol. Spectrosc.* **2012**, *88*, 171–176. [CrossRef]
7. Ma, H.; Zhou, J.; Hua, L.; Cheng, F.; Zhou, L.; Qiao, X. Chromium recovery from tannery sludge by bioleaching and its reuse in tanning process. *J. Clean. Prod.* **2017**, *142*, 2752–2760. [CrossRef]
8. Dong, H.; Xuguang, J.; Lv, G.; Yong, C.; Jianhua, Y. Co-combustion of tannery sludge in a commercial circulating fluidized bed boiler. *Waste Manag.* **2015**, *46*, 227–233. [CrossRef]
9. Abbas, N.; Jamil, N.; Hussain, N. Assessment of key parameters in tannery sludge management: A prerequisite for energy recovery. *Energy Sources Part A Recover. Util. Environ. Eff.* **2016**, *38*, 2656–2663. [CrossRef]
10. Skierka, E.; Sadowska, M. The influence of different acids and pepsin on the extractability of collagen from the skin of Baltic cod (Gadus morhua). *Food Chem.* **2007**, *105*, 1302–1306. [CrossRef]
11. Ferreira, M.; Almeida, M.; Pinho, S.; Santos, I. Finished leather waste chromium acid extraction and anaerobic biodegradation of the products. *Waste Manag.* **2010**, *30*, 1091–1100. [CrossRef]
12. Chronska, K.; Przepiorkowska, A. Buffing dust as a filler of carboxylated butadiene-acrylonitrile rubber and butadiene-acrylonitrile rubber. *J. Hazard. Mater.* **2007**, *151*, 348–355. [CrossRef]
13. Chrońska-Olszewska, K.; Przepiórkowska, A. A Mixture of Buffing Dust and Chrome Shavings as a Filler for Nitrile Rubbers. *J. Appl. Polym. Sci.* **2011**, *122*, 2899–2906. [CrossRef]
14. Przepiórkowska, A.; Chjrońska, K.; Zaborski, M. Chrome-tanned leather shavings as a filler of butadiene–acrylonitrile rubber. *J. Hazard. Mater.* **2007**, *141*, 252–257. [CrossRef] [PubMed]
15. Tadeusiak, A.; Przepiórkowska, A. Próba wykorzystania odpadów skór garbowanych do uzyskania biorozkładalnych materiałów białkowo-elastomerowych. *Przetwórstwo tworzyw.* **2011**, *5*, 393–396.
16. Przepiórkowska, A.; Prochoń, M.; Zaborski, M.; Żakowska, Z.; Piotrowska, M. Biodegradable protein-containing elastomeric vulcanizates. *Rubber Chem. Technol.* **2005**, *75*, 868–878. [CrossRef]

17. Ravichandran, K.; Natchimuthu, N. Vulcanization characteristics and mechanical properties of natural rubber-scrap rubber compositions filled with leather particles. *Polym. Int.* **2005**, *54*, 553–559. [CrossRef]
18. Rethinam Senthil, R.; Hemalatha, T.; Manikandan, R.; Nath Das, B. Leather boards from buffing dust: A novel perspective. *Clean. Technol. Environ. Policy* **2015**, *17*, 571–576. [CrossRef]
19. Rajaram, J.; Rajnikanth, B.; Gnanamani, A. Preparation characterization and application of leather particulate-polymer composites (LPPCs). *J. Polym. Environ.* **2009**, *17*, 181–186. [CrossRef]
20. Kowalska, E.; Żubrowiska, M. Termoplastyczne kompozyty napełniane odpadami poprodukcyjnymi z garbowania skóry. *Recykling. Przegląd Komunal.* **2004**, *2*, 14–15.
21. Zhou, J.; Yu, J.; Liao, H.; Zhang, Y.; Luo, X. Facile fabrication of bimetallic collagen fiber particles via immobilizing zirconium on chrome-tanned leather as adsorbent for fluoride removal from ground water near hot spring. *Sep. Sci. Technol.* **2020**, *4*, 658–671. [CrossRef]
22. Arcibar-Orozco, J.A.; Barajas-Elias, B.S.; Caballero-Briones, F.; Nielsen, L.; Rangel-Mendez, J.R. Hybrid Carbon Nanochromium Composites Prepared from Chrome-Tanned Leather Shavings for Dye Adsorption. *Water Air Soil Pollut.* **2019**, *230*, 1–13. [CrossRef]
23. Kamaraj, C.; Lakshmi, S.; Rose, C.; Muralidharan, C. Wet Blue Fiber and Lime from Leather Industry Solid Waste as Stabilizing Additive and Filler in Design of Stone Matrix Asphalt. *Asian J. Res. Soc. Sci. Humanit.* **2017**, *7*, 240–257. [CrossRef]
24. Ma, F.; Dinga, S.; Renb, H.; Penga, P. Preparation of chrome-tanned leather shaving-based hierarchical porous carbon and its capacitance properties. *RSC Adv.* **2019**, *9*, 18333–18343. [CrossRef]
25. Liu, y.; Wang, Q.; Li, L. Reuse of leather shavings as a reinforcing filler for poly (vinyl alcohol). *J. Thermoplast. Compos. Mater.* **2014**, *29*, 1–15.
26. Şaşmaz, S.L.; Karaağaç, B.; Uyanık, N. Utilization of chrome-tanned leather wastes in natural rubber and styrene-butadiene rubber blends. *J. Mater. Cycles Waste Manag.* **2019**, *21*, 166–175. [CrossRef]
27. Polgar, L.M.; Kingma, A.; Roelfs, M.; van Essen, M.; van Duin, M.; Picchioni, F. Kinetics of cross-linking and de-cross-linking of EPM rubber with thermoreversible. *Diels-Alder Chem. Eur. Polym. J.* **2017**, *90*, 150–161.
28. Prochoń, M.; Marzec, A.; Szadkowski, B. Preparation and Characterization of New Environmentally Friendly Starch-Cellulose Materials Modified with Casein or Gelatin for Agricultural Applications. *Materials* **2019**, *12*, 1–19.
29. Baeta, D.A.; Zattera, J.A.; Oliveira, M.G.; Oliveira, P.J. The use of styrene-butadiene rubber waste as a potential filler in nitrile rubber: Order of addition and size of waste particles. *J. Chem. Eng.* **2009**, *26*, 23–31. [CrossRef]
30. Kium, K.-H.; Lee, J.-Y.; Choi, J.-M.; Kim, H.J.; Seo, B.; Kim, B.-S.; Kwag, G.; Paik, H.-J.; Kim, W. Synthesis of Ionic Elastomer Based on Styrene-Butadiene Rubber Containing Methacrylic Acid. *Elastomer Compos.* **2013**, *48*, 46–54. [CrossRef]
31. Brandt, H.-D.; Nentwig, W.; Rooney, N.; LaFlair, R.T.; Wolf, U.U.; Duffy, J.; Puskas, J.E.; Kaszas, G.; Drewitt, M.; Glander, S. *Rubber, 5. Solution Rubbers*; Encyclopedia of Industrial Chemistry Wiley-VCH: Hoboken, NJ, USA, 2012.
32. Prochoń, M.; Przepiórkowska, A. Innovative application of biopolymer keratin as a filler of synthetic acrylonitrile-tadiene rubber NBR. *J. Chem.* **2013**, *2013*, 1–8. [CrossRef]

© 2020 by the authors. Licensee MDPI, Basel, Switzerland. This article is an open access article distributed under the terms and conditions of the Creative Commons Attribution (CC BY) license (http://creativecommons.org/licenses/by/4.0/).

Article

Influence of Bio-Based Plasticizers on the Properties of NBR Materials

Md Mahbubur Rahman [1,*], Katja Oßwald [2], Katrin Reincke [2] and Beate Langer [1,2]

[1] Department of Engineering and Natural Sciences, Hochschule Merseburg-University of Applied Sciences, Eberhard-Leibnitz-Straße 2, 06217 Merseburg, Germany; beate.langer@hs-merseburg.de
[2] Polymer Service GmbH Merseburg, Associate institute of Hochschule Merseburg-University of Applied Sciences, 06217 Merseburg, Germany; katja.osswald@psm-merseburg.de (K.O.); katrin.reincke@psm-merseburg.de (K.R.)
* Correspondence: md_mahbubur.rahman@hs-merseburg.de; Tel.: +49-3461-462717

Received: 19 March 2020; Accepted: 27 April 2020; Published: 1 May 2020

Abstract: A high number of technical elastomer products contain plasticizers for tailoring material properties. Some additives used as plasticizers pose a health risk or have inadequate material properties. Therefore, research is going on in this field to find sustainable alternatives for conventional plasticizers. In this paper, two modified bio-based plasticizers (epoxidized esters of glycerol formal from soybean and canola oil) are of main interest. The study aimed to determine the influence of these sustainable plasticizers on the properties of acrylonitrile–butadiene rubber (NBR). For comparison, the influence of conventional plasticizers, e.g., treated distillate aromatic extract (TDAE) and Mesamoll® were additionally investigated. Two types of NBR with different ratios of monomers formed the polymeric basis of the prepared elastomers. The variation of the monomer ratio results in different polarities, and therefore, compatibility between the NBR and plasticizers should be influenced. The mechanical characteristics were investigated. In parallel, dynamic mechanical analysis (DMA) and thermogravimetric analysis (TGA) were performed and filler macro-dispersion was determined. Bio-based plasticizers were shown to have better mechanical and thermal properties compared to conventional plasticizers. Further, thermo-oxidative aging was realized for 500 h, and afterwards, mechanical characterizations were done. It was observed that bio-based plasticizers have almost the same aging properties compared to conventional plasticizers.

Keywords: bio-oil; bio-based plasticizer; eco-friendly plasticizer; acrylonitrile-butadiene rubber; NBR; mechanical testing; thermo-oxidative aging

1. Introduction

Acrylonitrile-butadiene rubber (NBR) is a synthetic unsaturated statistical copolymer of acrylonitrile (ACN) and butadiene. NBR has good oil and chemical resistance [1]. It is used in the automotive and petroleum industries for oil and engine fuel transport equipment, for machinery pumps and in disposable nonlatex gloves [2]. It is a perfect choice for sealing applications due to its resistive properties, e.g., as O-rings due to its oil resistance [3]. Excellent processability is also a concern. Despite application, the field varies with the proportion of ACN content and molecular weight of the polymer. It has excellent resistance to petroleum products over a wide temperature range [3]. NBR typically contains extender oil that works as a plasticizer. Unfilled NBR has a low level of mechanical properties; thus, filler is incorporated into the NBR. This increases the viscosity and reduces the internal plasticization, so the addition of an external plasticizer is needed [4]. Various types of oil can be added to NBR compounds to reduce viscosity, improve processing properties, increase low-temperature flexibility, and reduce production costs [5].

The International Union of Pure and Applied Chemistry (IUPAC) developed a universally accepted definition of plasticizers in 1951: "A plasticizer is a substance or material incorporated in a material to increase its flexibility, workability, or distensibility" [6]. The nonrenewable resource mineral oil is mostly used in different countries and from which aromatic, naphthenic, and paraffinic oils are produced. Among those, aromatic oil is a very compatible oil with NBR due to its polar properties [7].

However, in 1994 a report by Swedish National Chemicals showed that polycyclic aromatic hydrocarbon (PCA) is the main constituent of aromatic oil, which has been proved harmful for the environment and human health [8].Aromatic process oils (distillate aromatic extracts, DAE)are classified as carcinogenic to humans [9]. Due to treatment, the aromatic content is reduced and these products (treated distillate aromatic extracts, TDAE) are state of the art. TDAE contains 1.3 wt % PCA and according to European legislation, the maximum content of PCA should be 3 wt % [9].

Several researchers have been looking for a sustainable replacement for conventional petroleum oils in the rubber industry. Vegetable oil, such as soybean oil [5], linseed oil [10], castor oil [11], coconut oil [4], and rice bran oil [12], is a renewable and inexpensive oil resource, and research aims to substitute conventional petroleum oils with vegetable oils.

Plasticizers have a specific solubility in rubber and contribute to the Brownian motion of polymer chains; therefore, they reduce the viscosity of the rubber compound [5]. Usually, no chemical reactions take place during the compounding process. The mixing temperature and the kneader-screw-rotation speed influence the mixing properties of the rubber during the mixing process. The temperature of vulcanization of sulfur occurs between 150 °C and 180 °C [13].

The choice of a plasticizer must be matched to the polymer; thus, nonpolar plasticizers are generally used for nonpolar polymers, whereas polar polymers require polar plasticizers [5]. The polar property of NBR thus influences the selection of a plasticizer. The latter is controlled by the ACN content in the NBR backbone chain. Ordinary natural fatty oil is slightly polar so that it becomes ready to react with some active parts to form fatty oil derivatives or to polymerize fatty oils. Moreover, since fatty oils are polar, they are very compatible with NBR (high ACN content). The epoxidized fatty oil contains fewer double bonds compared to conventional fatty oil and contains the active oxirane group. A plasticizer containing OH groups is most compatible with polar polymers, such as NBR [14]. The active sites are available to realize a dipole–dipole interaction with NBR, although the hydrogen bond and Van der Waals bond influences the compatibility. Research was conducted on the plasticization mechanism in the 20th century, and several theories have been postulated on the plasticization mechanism [14–17].

Khalaf et al. [18] worked with selected vegetable oils as plasticizers for NBR elastomer. They used olive oil and orange oil. The motivation was to find are placement for conventional oil, such as dioctyl phthalate (DOP). Zhu et al. [19] found that the mechanical properties were significantly enhanced by the addition of vegetable oils. Wang et al. [20] used palm oil as the source of renewable plasticizer with ethylene-propylene-diene rubber (EPDM). The investigation revealed that palm oil reduces the Mooney viscosity as well as increases some of the selected mechanical properties. Pechurai et al. [21] worked on castor oil and jatropha oil with styrene-butadiene rubber (SBR). Pechurai showed that the addition of vegetable oils profoundly enhanced the mechanical properties of SBR, and vegetable oils were the correct replacement of petroleum-based oil. Xuan Liu et al. [22] worked on the thermo-oxidative aging of NBR. In their investigation, the authors studied the migration of low-molecular additives as well as chemical changes like post-crosslinking of the materials by using different analytical tests. It was found that during oven aging, post-curing occurred due to high temperatures.

In the present study, epoxidized esters of glycerol formal from soybean and canola oil were used as sustainable, bio-based plasticizers. Two conventional plasticizers, TDAE and Mesamoll®, were applied to compare the characteristics. The main components of the vegetable oils are fatty acid esters. The fatty acid contains several double bonds in the fatty acid chains. The epoxidized groups are more active than the double bonds in the fatty acid chains [23]. Thus, it is anticipated that the modified fatty oil is more active than unmodified fatty oil. The active double bonds can play an essential role during the curing of the NBR compounds.

The curing behavior was studied as well as the thermal, mechanical, and dynamic mechanical properties. To investigate the NBR/CB/plasticizer particle dispersion in the NBR compound, an optical analysis was done. Additionally, for selected materials, thermo-oxidative aging was performed.

2. Materials and Methods

The investigations were performed on NBR with two different contents of ACN monomer. Arlanexo Deutschland GmbH (Dormagen, Germany) supplied Perbunan® 3445F (34% ACN) and Perbunan®1846F (18% ACN). The filler was CORAX® N 550 carbon black (CB) from Orion Engineered carbons GmbH (Frankfurt/M., Germany). TDAE was used from Hansen & Rosenthal KG (Hamburg, Germany). Mesamoll® was obtained from Lanxess Deutschland GmbH (Leverkusen, Germany) and bio-oils, and epoxidized ester of glycerol formal from soybean oil (EESO) and epoxidized ester of glycerol formal from canola oil (EECO) from Glaconchemie GmbH (Merseburg, Germany) were used.

2.1. Processing of Plasticized Compounds

NBR with 34% ACN and 18% ACN were named NBR-34 and NBR-18 accordingly. The formulation of the materials, which is reported in Table 1, differs mainly in the NBR type, plasticizer types, and contents, whereas the other mixing ingredients were kept almost unchanged.

Table 1. Composition and time of addition during mixing.

Name	Comments	Content (phr)	Time of Addition to the Kneader (min)
NBR	34% ACN, 18% ACN	100	0
Carbon black	N 550	40	1
Oil	TDAE, Mesamoll®, EESO, EECO	0, 5, 15	1
Stearic acid	Processing aids	1	1
ZnO	Activator	3	1
6PPD[1]	Antioxidant	1.5	1
Sulfur	Crosslinking agent	1.75	5
CBS[2]	Accelerator	1.05	5

[1] 6PPD, N-(1,3-dimethylbutyl)–N'–phenyl–p–phenylenediamine; [2] CBS, N–cyclohexyl–benzothiazole–2–sulfonamide

The compounds were prepared with a two-stage mixing procedure in a lab kneader with a fill factor of 0.7. The starting temperature in the kneader was 50 °C, and a rotation speed of 50 rpm was used. The polymer was introduced first in the chamber, and the CB, plasticizer, stearic acid, ZnO, and 6PPD were added after 1 min. After 5 min, the sulfur and CBS were added. The compounded stock was discharged after 10 min, and as a final step, it was homogenized on a two-roll mill before curing.

Further, cure characteristics at a temperature of 160 °C were determined by using a vulcameter (type GÖTTFERT elastograph, GÖTTFERT Werkstoff-Prüfmaschinen GmbH, Buchen, Germany). According to the results of the vulcameter tests, the time t_{90} was fixed as the vulcanization time for the necessary plates. For each material (formulation), plates with a size of 120 mm× 120 mm × 2 mm were vulcanized. From these plates, specimens of the desired form for the tests were cut using a metal cutter.

2.2. Tests

Hardness (Shore A) of the samples was determined as per DIN ISO 7619 standard [24], using a Zwick hardness tester (type 7206.H04, ZwickRoell, Ulm, Germany). Tensile and tear strength were determined according to ISO 37 [25] and ISO 34-1 [25], respectively, using the universal testing machine (type Zwick Z020, ZwickRoell, Ulm, Germany).

Compression set (CS) values were determined according to DIN ISO 815-1 [25]. Samples with 6.3 mm thickness and 13 mm diameter were compressed to constant strain (~25%) and kept in this condition for 24 h at room temperature. Finally, the CS in % was calculated.

$$\text{CS (\%)} = \frac{h_0 - h_i}{h_0 - h_s} \times 100 \tag{1}$$

Here, h_0 is the thickness of the sample before compression; h_i is the thickness of the sample after recovery, and h_s is the height in mm of the spacer.

The dynamic mechanical analysis (DMA) was carried out on a GABO DMA equipment (type Eplexor 500 N, GABO QUALIMETER Testanlagen GmbH, Ahlden, Germany), according to standard ISO 6721-4 [26]. The storage modulus E' and the loss modulus E'', as well as the mechanical loss factor tan δ, were determined in dependence on the temperature, and the glass transition temperature T_g was obtained from these data.

Thermogravimetric analysis of the samples was carried out on a Mettler Toledo instrument with a heating rate of 20 K/min under anitrogen atmosphere up to 600 °C. Then, the atmosphere was changed to oxygen. The following characteristics are determined from the thermograms: the temperature of onset of degradation and the temperature at the peak rate of decomposition, the peak rate of degradation, and the weight of residue remaining at 600 °C. According to ASTM D 2663 [27], the macro filler dispersion was determined. A dispersion index of 100% means that no agglomerate size larger than 6 μm could be found on the cut surface [28]. The samples were prepared by cutting with a sharp metal razor and then investigated with a Leica DM 2700M light microscope.

For the aging test, samples were kept for 500 h at 23°C, 40 °C, and 80 °C. Finally, changes in material behavior were characterized by tensile, hardness, and compression set testing.

3. Results

The cure characteristics were examined by using a moving disc rheometer, and the curing curves acquired from NBR-34 compounds with 15 phr of different plasticizers areshown exemplarily in Figure 1. It was clearly evident that, due to the dilution effects [20] of the plasticizers, the maximum torque M_{max} values decreased when the plasticizers were simply added. M_{max} is reduced gradually up to 5phr, and a significant reduction has occurred at 15 phr. Minimum torque M_{min} is comparable, independent of the type of plasticizer, as is shown in Figure 1 for NBR-34.

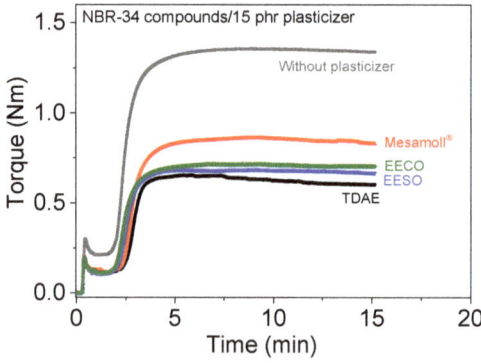

Figure 1. Vulcanization curves of various plasticized NBR compounds.

Figure 2 gives the torque differences ($\Delta M = M_{max} - M_{min}$) of NBR vulcanizate as a function of plasticizer content. The lower ΔM indicates one of the possible reasons for the low degree of crosslinking of compounds with a higher loading of plasticizers [4,11,29]. ΔM for plasticizer-loaded

NBR-18 is comparatively more significant than for NBR-34, as shown in Figure 2. A large number of unsaturated bonds in NBR-18 due to the butadiene monomer could be responsible for the higher degree of crosslinking [30].

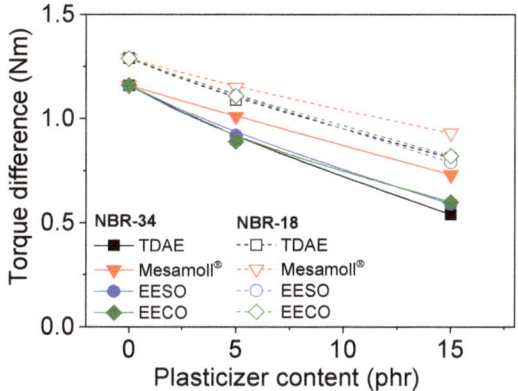

Figure 2. Torque differences during curing of plasticized NBR compounds based on NBR-34 and NBR-18 as a function of plasticizer content.

The vulcanization time of NBR compounds is shown in Figure 3. The NBR-34 vulcanizates clearly showed a lower curing time compared to the NBR-18 as well as lower difference in torque values (see Figure 2). For NBR-34, the addition of plasticizer changes the vulcanization time, and there is minimal difference between the four plasticizers. TDAE- and Mesamoll®-loaded NBR-18 showed increased cure time. However, EECO decreases the vulcanization time of NBR-18. The optimum conversion of epoxidation of bio-based plasticizers is about 90% [31]. As a result, bio-based plasticizers contain 10% fatty acid ester. The acid value of EECO is about 2.32 mg of KOH/mg, whereas the acid value of EESO is about 0.68 KOH/mg [7]. The higher acid value corresponds to the higher free fatty acid. The fatty acid acts as a co-activator during the formulation of NBR compounds [32].

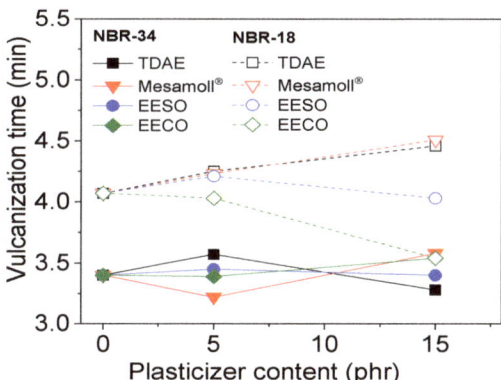

Figure 3. Vulcanization time as a function of plasticizer content for the investigated NBR compounds.

Figure 4 gives the variation of the tensile strength of the NBR vulcanizates with plasticizer loading. For most compounds, the tensile strength is slightly decreased or remains nearly constant with increasing plasticizer content. If 5 phr Mesamoll® and EECO, respectively, are added, the tensile strength increases. This may contribute to more homogenous filler dispersion in the rubber matrix, as reported in [4]. However, our own examinations of the macro-dispersion index gave no correlation with the tensile strength values for increasing amount of plasticizer. The tensile strength of bio-based plasticized NBR-34 is slightly higher compared to NBR-18. One possible reason for this is the higher polarity of NBR-34 due to higher content of ACN compared to NBR-18 [33]. The conventional plasticizers do not contain any OH groups, and therefore, compatibility with a polar polymer-like NBR [14] is restricted, leading to a lower tensile strength.

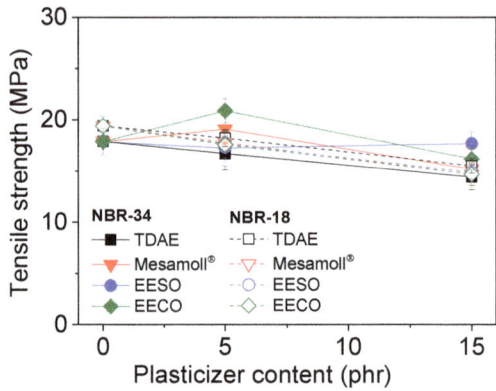

Figure 4. Variation of tensile strength with plasticizer content for NBR vulcanizates.

Generally, a decrease in tensile strength is often combined with an increase in the strain at break [34]. In principle, this is also the case for the materials investigated in this study, as can be seen in Figure 5 showing strain-at-break values of the NBR vulcanizates with different loading of plasticizer. There is a general trend of increasing deformability with rising amount of plasticizer. In detail, some differences can be seen for the different plasticizers. For the NBR-18 compounds, the strain at break remains constant within standard deviation when adding 5 phr of the plasticizers, with the exception of EESO. This may be due to not having optimal filler dispersion or the antiplasticization effect [35].The NBR-34 compounds with bio-based plasticizers have a higher strain at break compared to the compounds with TDAE and Mesamoll®. Again, the different polarity of NBR-18 may be the reason for the lower deformability of the materials with the bio-based plasticizers, compared to NBR-34. From the results, a more substantial lubrication effect of the bio-based plasticizers might be derived; in other words, bio-based plasticizers might promote more pronounced polymer chain motion in the vulcanizates [20]. Another reason for the different level of the mechanical properties depending on plasticizer type and amount may be a possible influence on the crosslink density. This was not investigated explicitly, but from the vulcameter curves in Figure 1 it is seen that the maximum torques of the compounds vary, depending on the used plasticizer. This can be a sign of a varying crosslink density. With the currently available data, a chemical reaction between plasticizer and the vulcanization system cannot be excluded, which could lead to a lower crosslink density.

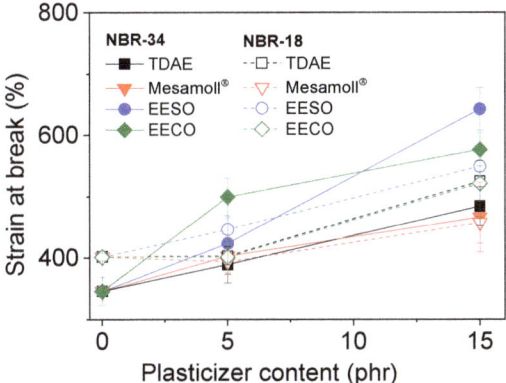

Figure 5. Variation of strain at the break with plasticizer content for NBR vulcanizates.

The tear strength values (see Figure 6) show a significant increase in the case of bio-based plasticizers, especially at higher plasticizer loading. In the case of TDAE and Mesamoll®, the tear strength remains almost constant at around 8 N/mm. In contrast, the bio-based plasticized NBR compounds showed an increase of up to 9 to 11 N/mm at 15 phr loading range. Tearing of NBR vulcanizate occurs due to the propagation of cracks initiated at the stress concentration point through the wearing of rubber molecules at the NBR–carbon black interfaces [10]. The micro-plasticization of the interfaces using an effective plasticizer can hinder the propagation of cracks [36]. The tear strength is not much influenced by increasing amount of conventional plasticizers.

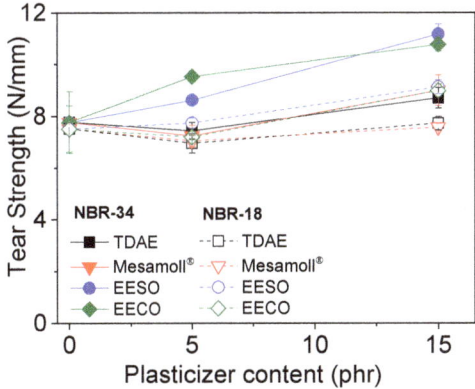

Figure 6. Dependence of tear strength on plasticizer content for NBR vulcanizates.

Figure 7 shows the variation of hardness with plasticizer concentration. Generally, the hardness is higher without the addition of any plasticizer. With the plasticizers, the materials show continuously decreasing hardness for both types of NBR vulcanizates when increasing the plasticizer content. However, NBR-18 shows lower hardness compared to NBR-34. For both NBR vulcanizates with Mesamoll®, highest hardness values were obtained compared to other systems. This is perhaps because of the active participation of Mesamoll® during the vulcanization of NBR. Kukreja [29] stated that the highest hardness might be attributed due to the high participation of crosslinking.

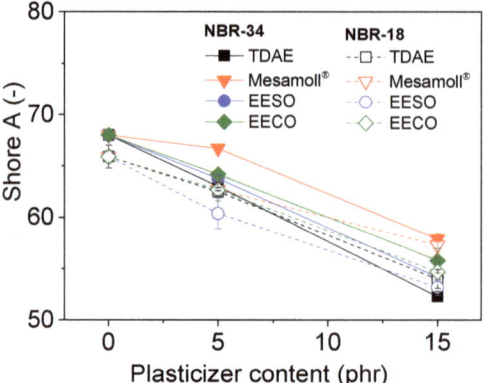

Figure 7. Shore A hardness values as a function of plasticizer content for NBR vulcanizates.

Figure 8 shows the CS values of different oil content of two NBR vulcanizates. Both types of NBR with bio-based plasticizers have a higher CS, which increases with oil content. The use of Mesamoll® and TDAE results in a more or less unchanged CS. A previous study [4] proved that in a rubber vulcanizate with low oil content, the filler dispersion was poor, but at higher oil content, the plasticizing effect and segmental mobility were pronounced, leading to higher CS values [4]. In a previous study [11], compression set values correlated to the hardness values. The highest CS was found for the material with the lowest hardness. Here, the same effect could be found. A low CS value represents a better recovery behavior after load release. However, the bio-based plasticizers show a maximum CS value of ~12%; this is still a low value compared to the threshold, which is ~40% for a gasket [37].

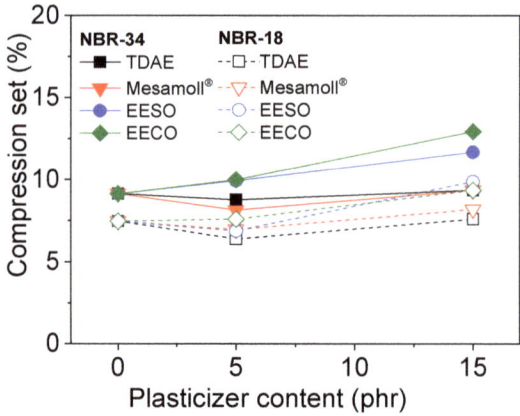

Figure 8. Compression set (CS) values as a function of plasticizer content for NBR vulcanizates.

The temperature dependencies of storage modulus E' of the plasticizer-loaded NBR vulcanizates are shown in Figure 9. For the values of E' in the glassy state all the NBR vulcanizates are similar to each other. In the high-temperature region corresponding to the rubbery state, the NBR vulcanizate without any plasticizer has the highest value of E', followed by NBR/EECO, NBR/Mesamoll®, NBR/EESO, and NBR/TDAE vulcanizates (see Table 2). Figure 10 shows the temperature dependences of loss modulus E'' of various plasticizer-loaded NBR vulcanizates. The E'' values of all NBR vulcanizates in the glassy state are similar to each other. In the rubbery state, the NBR vulcanizate without any plasticizer shows the highest value. When using plasticizers, E' and E'' values are decreased. The E'' values of bio-based plasticizer-loaded NBR vulcanizate are slightly higher (see Table 2). The storage modulus E' decreases with increasing temperature, also after passing the glass transition temperature; this means it is in the entropy-elastic range. This is a typical behavior of carbon-black-filled elastomers and has importance for the use of the material at higher temperatures, e.g., in the case of sealants, where a certain stiffness is necessary.

Table 2. Dynamic mechanical analysis (DMA) analysis of NBR-34 vulcanizates with various plasticizers.

NBR-34	tan δ	T_g(°C)	E' (MPa) at 23 °C	E'' (MPa) at 23 °C
Without plasticizer	1.32	−6.88	11.53	2.31
15 phr TDAE	1.43	−7.34	6.98	1.44
15 phr Mesamoll®	1.40	−12.11	7.66	1.37
15 phr EESO	1.36	−14.11	7.55	1.47
15 phr EECO	1.34	−12.56	7.91	1.57

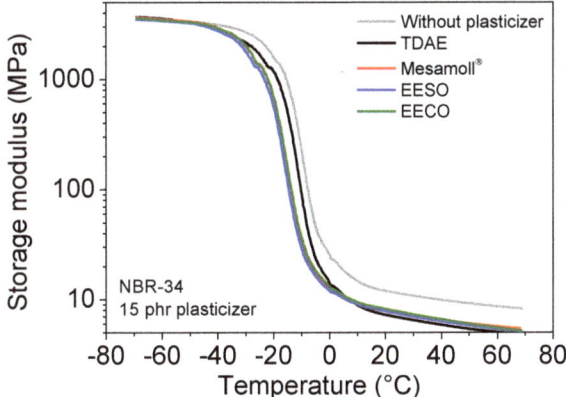

Figure 9. Storage modulus E' of NBR-34 vulcanizates with 15 phr various plasticizers as a function of temperature.

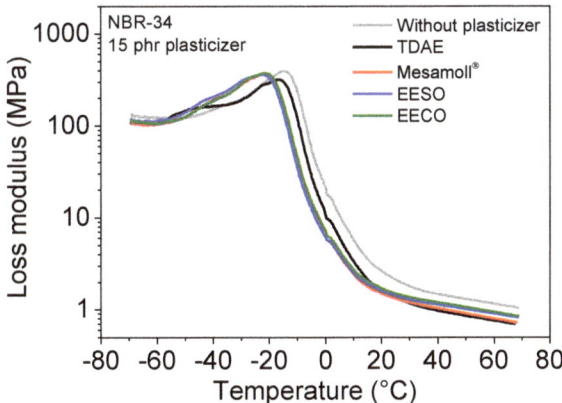

Figure 10. Loss modulus E'' of NBR-34 vulcanizates with 15 phr various plasticizers as a function of temperature.

Figure 11 shows that all the NBR vulcanizates have a single tan δ peak. The tan δ value converges to a similar value in the low-temperature region. In the high-temperature region, the NBR/bio-based plasticizers have higher values of tan δ compared to NBR/TDAE or NBR/Mesamoll®. According to previous studies [38,39], the values of tan δ and E' predicts rubber performance. The maximum of tan δ of the compounds is shifted to lower values in the case of the bio-based plasticizers EESO and EECO as well as the synthetic plasticizer Mesamoll®. This is due to the lower T_g values of these plasticizers. For practical application, a lower T_g can be of importance, because molecular mobility starts at lower temperatures. This means, the flexibility of the material is given in a larger temperature range. Further, because of the viscoelasticity, a mechanical material loading with higher frequencies can lead to a shift of T_g to higher temperatures. Therefore, materials with basically lower T_g are advantageous. The heights of the tan δ peak are not strongly different for the different plasticizers. This means, the energy loss in the network is comparable.

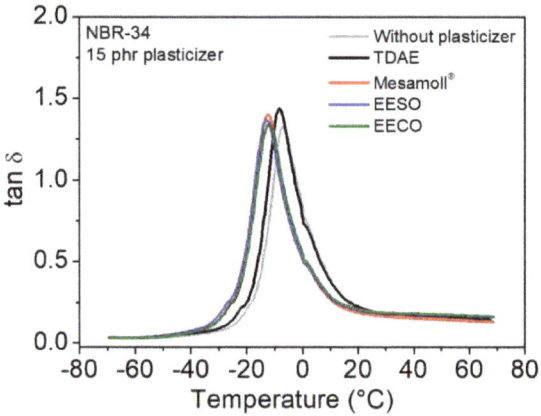

Figure 11. Mechanical loss factor tan δ of NBR-34 vulcanizates with 15 phr various plasticizers as a function of temperature.

The glass transition temperature T_g was pointed out in Table 2. T_g is highly influenced by the chemical structure and molecular weight of plasticizers. High molecular-weight plasticizer has high free volume, and high free volume decreases the T_g [19,20]. Here, the molecular weight of TDAE is around 180 g/mol, and for the bio-based plasticizers it is approximately 400 g/mol.

The thermal stability of plasticizer-loaded NBR vulcanizate was characterized using TGA as a function of various plasticizers and is shown in Figure 12, while the results are summarized in Table 3. NBR vulcanizate without plasticizer shows the best thermal stability. It is noticed that TDAE and bio-based plasticizer-loaded NBR vulcanizates show an almost similar decomposition process and this has improved the first decomposition stage in comparison to Mesamoll®; this is because of the ester groups in the bio-based plasticizers [40]. Generally, these plasticizer-loaded NBR vulcanizates show 3 main decomposition steps. Initially, the plasticizer and all other low-molecular components are decomposed up to ca. 350 °C. Then, the polymer is thermally decomposed in the range of 360–500 °C. According to [41], the initial decomposition of CB-reinforced NBR starts from 360 °C. Here, the samples were pyrolyzed to ~32%. After this second decomposition step, the pyrolysis carbon and the filler carbon black remain. The last decomposition step starts at 600 °C under the influence of oxygen and is completed at 650 °C. The decomposition stages are consistent with degradation due to random chain scission of the butadiene and the nature of the ACN parts in NBR vulcanizate [42]. In addition, there are very strong electronegative groups that result in relatively high interaction and high heat resistance in elastomers.

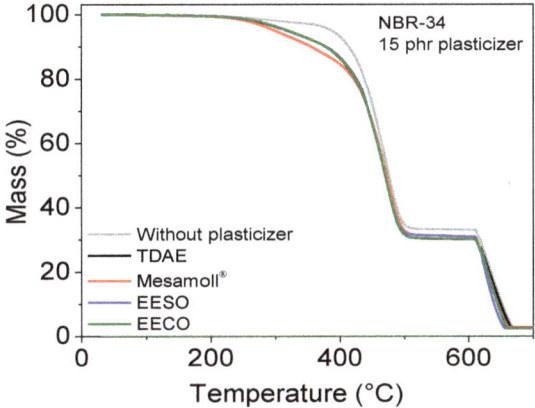

Figure 12. TGA thermograms of NBR-34 vulcanizates with various plasticizers.

Table 3. Results of TGA of NBR-34 vulcanizates with various plasticizers.

NBR-34	Decomposition % (250 °C)	Decomposition % (350 °C)	Decomposition % (450 °C)	Decomposition % (550 °C)
Without plasticizer	0.96	2.92	27.55	66.89
15 phr TDAE	1.28	7.4182	33.23	69.74
15 phr Mesamoll®	1.73	9.44	33.08	68.66
15 phr EESO	1.17	7.60	33.6	68.72
15 phr EECO	1.28	7.36	33.74	69.66

Polarized microscopy was performed as a helpful tool for better understanding the extent of the plasticizers that influenced the dispersion within CB-reinforced NBR vulcanizates. Figure 13 shows the optical pictures of the investigated cross section areas of plasticized NBR vulcanizates. Here, Figure 13a shows a higher number of agglomerates compared to Figure 13b, but the size of agglomerates is smaller. A previous study [43] stated that the plasticizer decreases the shear stress between the CB occluded by the polymer chain. When the polymer is mixed, the polymer molecules have to slip over each other, which is the reason for more agglomerates remaining [15]. Here, the optical observation for Mesamoll®- and EESO-loaded NBR vulcanizates indicate a more homogeneous appearance compared to TDAE. TDAE-loaded NBR vulcanizate is depicted in Figure 13c. When the plasticizers were not well dispersed among the matrix, then the number of agglomerates decreased. This phenomenon is called antiplasticization [14]. Figure 13d shows TDAE coagulates within the polymer matrix inhomogeneously near the edge.

Tensile strength after thermo-oxidative aging of plasticizer-loaded NBR vulcanizate is shown in Figure 14. Results show that there is not much influence on tensile strength after thermo-oxidative aging. Shore A hardness consistently increased with increasing aging temperatures (see Figure 15). Figure 16 shows that the compression set decreases with increasing aging time. The reason for changing the mechanical properties after thermo-oxidative aging is perhaps due to the high crosslink formation and oxidated skin, which results from oxygen uptake at the surface of the specimen [44] or post vulcanization [5,45]. Further, the migration of volatile content and plasticizer happens due to long aging time and especially at high temperatures [46]. If the materials are kept at high temperatures for a long time, post-curing could happen during this time [30].

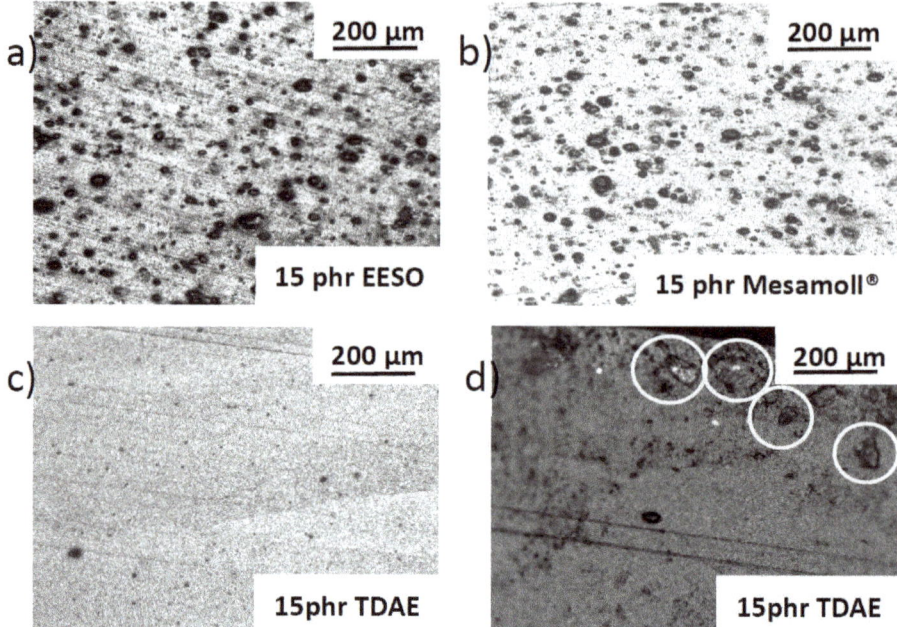

Figure 13. Cross-sectional view of different vulcanized rubber samples; (**a**) NBR-34 vulcanizate with 15 phr EESO, (**b**) NBR-34 vulcanizate with 15 phr Mesamoll®, (**c**) NBR-34 vulcanizate with 15 phr TDAE (middle of the samples), (**d**) NBR-34 vulcanizate with 15 phr TDAE (circles mark oil).

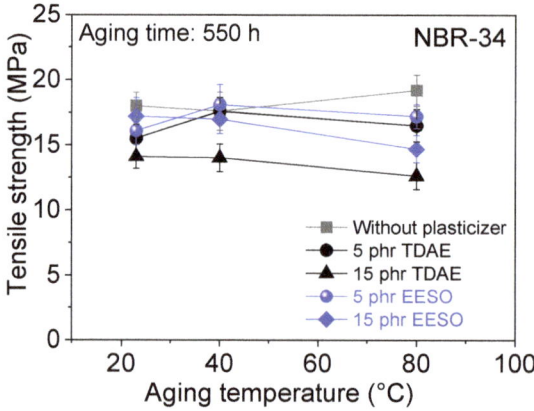

Figure 14. Dependence of tensile strength on aging temperature for NBR-34 vulcanizate with various plasticizers.

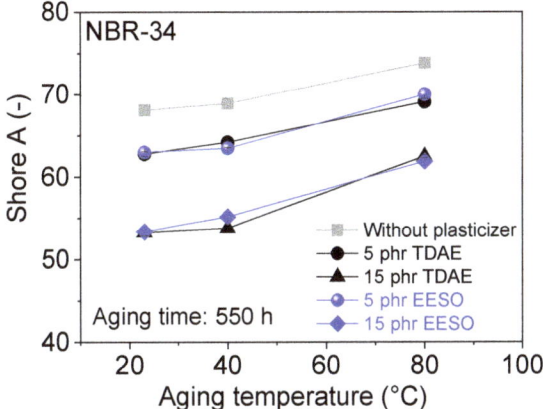

Figure 15. Shore A hardness as a function of aging temperature for NBR-34 vulcanizates with various plasticizers.

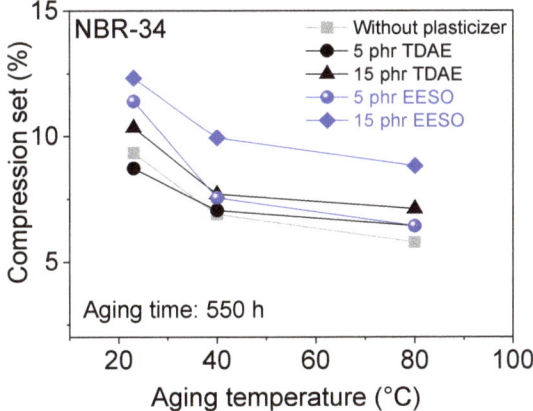

Figure 16. Compression set as a function of aging temperature for NBR-34 vulcanizates with various plasticizers.

4. Conclusions

The use of bio-based plasticizers (EESO, EECO), a conventional plasticizer (TDAE), and a synthetic plasticizer (Mesamoll®) in the compounding of NBR vulcanizates were studied. It was shown that bio-based plasticizers can advantageously be used as processing aids instead of conventional oils. Throughout the complete process chain, from preparation of the raw mixture and characterization of the kinetics of the crosslinking reaction to the final properties of the resulting elastomers, positive aspects of the use of bio-based plasticizers were observed, or at least a comparable level of the properties as for the use of mineral oil-based plasticizers. As an example, when using the bio-based plasticizers, there is a certain cure-accelerating effect, and also the crosslinking itself seems to be influenced. The mechanical properties are, generally, a result of the network development during vulcanization, and an improvement or at least a constancy in these properties is one of the main aims of material development. Based on the results shown in this paper, it can be concluded that for the investigated polymers, the use of such sustainable products is possible without noticeable loss in property levels. Future work will focus on other technical elastomers and on understanding of the influence of the plasticizer on the network development.

Author Contributions: Conceptualization, M.M.R. and K.R.; methodology, M.M.R. and K.O.; sample and specimen preparation, M.M.R.; experiments and analysis, M.M.R.; writing—original draft preparation, M.M.R.; writing—review and editing, M.M.R., K.R., K.O. and B.L.; supervision, K.O., K.R. and B.L.; project administration, K.R. and B.L.; funding acquisition, K.R. and B.L.; All authors have read and agreed to the published version of the manuscript.

Funding: This research was funded by the German Federal Ministry of Education and Research (BMBF) in the framework of the ZIM project: ZF4429401SL7.

Acknowledgments: The authors thank the Federal Ministry of Education and Research (BMBF) for financial support in the framework of the ZIM project: ZF4429401SL7. We would like to thank Glaconchemie GmbH for their bio-based plasticizer supply. Further, we thank Lanxess Deutschland GmbH for supplying the Mesamoll®.

Conflicts of Interest: The authors declare no conflict of interest. The funders had no role in the design of the study; in the collection, analyses, or interpretation of data; in the writing of the manuscript, or in the decision to publish the results.

References

1. Khalaf, A.; Yehia, A.; Ismail, M.N.; El-Sabbagh, S. High Performance Oil Resistant Rubber. *Open J. Org. Polym. Mater.* **2012**, *2*, 89–94. [CrossRef]
2. Hofmann, W. *Nitrile Rubber, Rubber Chemistry and Technology*, 1st ed.; Lancaster, P., Ed.; American Chemical Society: Lancaster, UK, 1964.
3. Waters, T.F. *Fundamentals of Manufacturing for Engineers*; CRC Press: London, UK, 1996; p. 96.
4. Raju, P.; Nandanan, V.; Kutty, S.K.N. A Study on the Use of Coconut Oil as Plasticiser in Natural Rubber Compounds. *J. Rubber Res.* **2007**, *10*, 1–16.
5. Petrović, Z.S.; Ionescu, M.; Milic, J.; Halladay, J.R. Soybean oil plasticizers as replacement of petroleum oil in rubber. *Rubber Chem. Technol.* **2013**, *86*, 233–249. [CrossRef]
6. Wilkes, C.D.C.; Summers, J. *PVC Handbook*; Hanser Publications: Munich, Germany; Cincinnati, OH, USA, 2005; pp. 172–193.
7. Dasgupta, S.; Agrawal, S.; Bandyopadhyay, S.; Chakraborty, S.; Mukhopadhyay, R.; Malkani, R.; Ameta, S. Characterization of eco-friendly processing aids for rubber compound. *Polym. Test.* **2007**, *26*, 489–500. [CrossRef]
8. Aisbl, E. *Replacement of Highly Aromatic Oils in Tyres*; European Tyre Rubber Manufacturers' Association: Brussels, Belgium, 2010; pp. 1–7.
9. Null, V. Safe process oils for tires with low environmental impact. *KGK-Kautsch. Gummi Kunstst.* **1999**, *52*, 799–805.
10. Fernandez, S.S.; Kunchandy, S.; Ghosh, S. Linseed Oil Plasticizer Based Natural Rubber/Expandable Graphite Vulcanizates: Synthesis and Characterizations. *J. Polym. Environ.* **2015**, *23*, 526–533. [CrossRef]
11. Raju, P.; Nandanan, V.; Kutty, S.K.N. A Study on the Use of Castor Oil as Plasticizer in Natural Rubber Compounds. *Prog. Rubber Plast. Recycl. Technol.* **2007**, *23*, 169–180. [CrossRef]
12. Kuriakose, A.P.; Rajendran, G. Use of rice-bran oil in the compounding of styrene butadiene rubber. *J. Mater. Sci.* **1995**, *30*, 2257–2262. [CrossRef]
13. Zhang, H.; Li, Y.; Shou, J.-Q.; Zhang, Z.-Y.; Zhao, G.-Z.; Liu, Y. Effect of curing temperature on properties of semi-efficient vulcanized natural rubber. *J. Elastomers Plast.* **2015**, *48*, 331–339. [CrossRef]
14. Wypych, G. Effect of plasticizers on properties of plasticized materials. In *Handbook of Plasticizers*, 2nd ed.; William Andrew Publications: Toronto, ON, Canada, 2012; pp. 209–306.
15. Kirkpatrick, A. Some Relations between Molecular Structure and Plasticizing Effect. *J. Appl. Phys.* **1940**, *11*, 255–261. [CrossRef]
16. Fox, T.G.; Flory, P.J. Second-Order Transition Temperatures and Related Properties of Polystyrene. I. Influence of Molecular Weight. *J. Appl. Phys.* **1950**, *21*, 581–591. [CrossRef]
17. Clark, F.W. Industrial and engineering chemistry. *Chem. Ind.* **1941**, *60*, 1089–1090.
18. Khalaf, A.; Ward, A.; El-Kader, A.E.A.; El-Sabbagh, S. Effect of selected vegetable oils on the properties of acrylonitrile-butadiene rubber vulcanizates. *Polimery* **2015**, *60*, 43–56. [CrossRef]
19. Zhu, L.; Cheung, C.S.; Zhang, W.; Huang, Z. Compatibility of different biodiesel composition with acrylonitrile butadiene rubber (NBR). *Fuel* **2015**, *158*, 288–292. [CrossRef]

20. Wang, Z.; Peng, Y.; Zhang, L.; Zhao, Y.; Vyzhimov, R.; Tan, T.; Fong, H. Investigation of Palm Oil as Green Plasticizer on the Processing and Mechanical Properties of Ethylene Propylene Diene Monomer Rubber. *Ind. Eng. Chem. Res.* **2016**, *55*, 2784–2789. [CrossRef]
21. Pechurai, W.; Chiangta, W.; Tharuen, P. Effect of Vegetable Oils as Processing Aids in SBR Compounds. *Macromol. Symp.* **2015**, *354*, 191–196. [CrossRef]
22. Liu, X.; Zhao, J.; Yang, R.; Iervolino, R.; Barbera, S. Effect of lubricating oil on thermal aging of nitrile rubber. *Polym. Degrad. Stab.* **2018**, *151*, 136–143. [CrossRef]
23. Zong, Z.; Soucek, M.D.; Liu, Y.; Hu, J. Cationic photopolymerization of epoxynorbornane linseed oils: The effect of diluents. *J. Polym. Sci. Part A Polym. Chem.* **2003**, *41*, 3440–3456. [CrossRef]
24. ISO 7619. *Rubber, Vulcanized or Thermoplastic—Determination Indentation Hardness Part 1 Durometer Method (Shore Hardness)*; International Organization for Standardization: Geneva, Switzerland, 2004.
25. DIN ISO 37:2011(E). *Rubber Vulcanized or Thermoplastic—Determiknation of Tensile Stress-Strain Properties*; International Organization for Standardization: Geneva, Switzerland, 2011.
26. ISO 6721. *(Plastics—Determination of Dynamic Mechanical Properties)*; International Organization for Standardization: Geneva, Switzerland, 2008; pp. 1–8.
27. ASTM D 2663. *Standard Test Methods for Carbon Black—Dispersion in Rubber 1*; ASTM International: West Conshohocken, PA, USA, 2000; pp. 1–6.
28. Le, H.; Ilisch, S.; Radusch, H.J. Characterization of the effect of the filler dispersion on the stress relaxation behavior of carbon black filled rubber composites. *Polymer* **2009**, *50*, 2294–2303. [CrossRef]
29. Kukreja, T.R.; Chauhan, R.; Choe, S.; Kundu, P.P. Effect of the doses and nature of vegetable oil on carbon black/rubber interactions: Studies on castor oil and other vegetable oils. *J. Appl. Polym. Sci.* **2003**, *87*, 1574–1578. [CrossRef]
30. Coran, A.Y. Vulcanization: Conventional and Dynamic. *Rubber Chem. Technol.* **1995**, *68*, 351–375. [CrossRef]
31. Petrović, Z.S.; Zlatanić, A.; Lava, C.C.; Sinadinović-Fišer, S. Epoxidation of soybean oil in toluene with peroxoacetic and peroxoformic acids—Kinetics and side reactions. *Eur. J. Lipid Sci. Technol.* **2002**, *104*, 293–299. [CrossRef]
32. Krejsa, M.R.; Koenig, J.L. A Review of Sulfur Crosslinking Fundamentals for Accelerated and Unaccelerated Vulcanization. *Rubber Chem. Technol.* **1993**, *66*, 376–410. [CrossRef]
33. Gong, L.; Nguyen, M.H.T.; Oh, E.-S. High polar polyacrylonitrile as a potential binder for negative electrodes in lithium ion batteries. *Electrochem. Commun.* **2013**, *29*, 45–47. [CrossRef]
34. Hassan, A.; Wahit, M.U.; Chee, C.Y. Mechanical and Morphological Properties of PP/LLPDE/NR Blends—Effects of Polyoctenamer. *Polym. Technol. Eng.* **2005**, *44*, 1245–1256. [CrossRef]
35. Kern Sears, J.; Darby, J., Jr. *The Technology of Plasticizers*; John Wiley Sons: New York, NY, USA, 1982; Volume 20, p. 459.
36. Kukreja, T.; Kumar, D.; Prasad, K.; Chauhan, R.; Choe, S.; Kundu, P.P. Optimisation of physical and mechanical properties of rubber compounds by response surface methodology—Two component modelling using vegetable oil and carbon black. *Eur. Polym. J.* **2002**, *38*, 1417–1422. [CrossRef]
37. McKeen, L.W. 1—Introduction to Creep, Polymers, Plastics and Elastomers. In *The Effect of Creep and Other Time Related Factors on Plastics and Elastomers*, 3rd ed.; McKeen, L.W., Ed.; William Andrew Publishing: Boston, MA, USA, 2015; pp. 1–41.
38. Schuring, D.J.; Futamura, S. Rolling Loss of Pneumatic Highway Tires in the Eighties. *Rubber Chem. Technol.* **1990**, *63*, 315–367. [CrossRef]
39. Flanigan, C.; Beyer, L.; Klekamp, D.; Rohweder, D.; Haakenson, D. Using bio-based plasticizers, alternative rubber. *Br. J. Psychiatry* **2013**, *2*, 15–19.
40. Wang, Z.; Zhang, X.; Wang, R.; Kang, H.; Qiao, B.; Ma, J.; Zhao, X.; Wang, H. Synthesis and Characterization of Novel Soybean-Oil-Based Elastomers with Favorable Processability and Tunable Properties. *Macromolecules* **2012**, *45*, 9010–9019. [CrossRef]
41. Alneamah, M.; Almaamori, M. Study of Thermal Stability of Nitrile Rubber/Polyimide Compounds. *Int. J. Mater. Chem.* **2015**, *5*, 1–3.
42. Doma, A.S.; Kamoun, E.A.; Abboudy, S.; Belal, M.A.; Khattab, S.N.; A El-Bardan, A. Compatibilization of Vulcanized SBR/NBR Blends using Cis-Polybutadiene Rubber: Influence of Blend Ratio on Elastomer Properties. *Eur. J. Eng. Res. Sci.* **2019**, *3*, 135–143. [CrossRef]

43. David, F. Cadogan, Phthalic acid and Derivates. In *Ullmann's Encyclopedia of Industrial Chemistry*; ICI Chemicals and Polymers: Cheshire, UK, 2012; pp. 35–154.
44. Brown, R.; Soulagnet, G. Microhardness profiles on aged rubber compounds. *Polym. Test.* **2001**, *20*, 295–303. [CrossRef]
45. Kundu, P.P.; Kukreja, T.R. Surface modification of carbon black by vegetable oil? Its effect on the rheometric, hardness, abrasion, rebound resilience, tensile, tear, and adhesion properties. *J. Appl. Polym. Sci.* **2002**, *84*, 256–260. [CrossRef]
46. Budrugeac, P. Accelerated thermal ageing of nitrile-butadiene rubber under air pressure. *Polym. Degrad. Stab.* **1995**, *47*, 129–132. [CrossRef]

© 2020 by the authors. Licensee MDPI, Basel, Switzerland. This article is an open access article distributed under the terms and conditions of the Creative Commons Attribution (CC BY) license (http://creativecommons.org/licenses/by/4.0/).

Article

Study of Carbon Black Types in SBR Rubber: Mechanical and Vibration Damping Properties

Marek Pöschl [1,2], Martin Vašina [2,3,*], Petr Zádrapa [1,2], Dagmar Měřínská [2] and Milan Žaludek [2]

[1] Centre of Polymer Systems, University Institute, Tomas Bata University in Zlin, Třída Tomáše Bati 5678, 760 01 Zlin, Czech Republic; poschl@utb.cz (M.P.); zadrapa@utb.cz (P.Z.)
[2] Faculty of Technology, Tomas Bata University in Zlin, Vavrečkova 275, 760 01 Zlin, Czech Republic; merinska@utb.cz (D.M.); zaludek@utb.cz (M.Ž.)
[3] Faculty of Mechanical Engineering, VŠB-Technical University of Ostrava, 17. Listopadu 15/2172, 708 33 Ostrava-Poruba, Czech Republic
* Correspondence: vasina@utb.cz

Received: 20 April 2020; Accepted: 19 May 2020; Published: 22 May 2020

Abstract: Styrene–butadiene rubber mixtures with four types of carbon black were studied in this paper. The mechanical properties, including the ability to damp mechanical vibration, were investigated, along with dynamical mechanical analysis (DMA). It has been found that carbon black types N 110 and N 330, having the largest specific surface area and the smallest particle diameter, provide a good stiffening effect. These particles have significant interactions between the rubber, resulting in good reinforcement. On the other hand, the carbon black N 990 type has a lower reinforcing effect and improved vibration damping properties at higher excitation frequencies due to higher dissipation of mechanical energy into heat under dynamic loading. The effect of the number of loading cycles on vibration damping properties of the rubber composites was also investigated in this study. It can be concluded that the abovementioned properties of the investigated rubber composites correspond to physical–mechanical properties of the applied carbon black types.

Keywords: vulcanized rubber; rubber compounds; carbon black; vibration damping; viscoelasticity

1. Introduction

Mechanical vibration, which is caused by oscillation of a mechanical or structural system about an equilibrium position, is an undesirable phenomenon in many cases (e.g., manufacturing processes, means of transport, and in home appliances). Furthermore, the mechanical vibration can contribute to excessive noise and have a negative effect on labor protection, manufacturing quality, and productivity. For these reasons, it is necessary to eliminate the mechanical vibration through appropriate measures, e.g., by application of suitable vibration damping materials [1–3].

Polymers have become more frequently used in a variety of applications in order to diminish vibration [2] in cars, ships, and other products where machinery creates vibration [4] or there are natural causes of vibration, which should be lowered from the environment in question. Anechoic chambers exist that allow for accurate observation of sound energy. Given or created energy is treated by dissipation or absorption by a variety of mechanisms [5].

In an engineering system, a structure should stay stable and undamaged despite internal and external vibrations. The stability of a system depends on its damping ability, which can be influenced by the composition of rubber mixtures [6,7]. Damping is defined as the energy dissipation of a vibration system and it can be presented by various parameters, one of which is the loss factor. The loss factor can be determined by several different methods, which are grouped into time, deformation, temperature, and frequency sweep tests [7,8].

All rubber mixture's properties are mainly influenced by its composition. The rubber type is only the base of the compound. In order to achieve the best properties in the finished rubber product, it is necessary to add some additives to the rubber to significantly change the properties of the product and improve its resistance to various undesirable effects. The additive component is given in units of phr, which is the weight per 100 parts by weight of rubber [9]. The most important rubber components include the vulcanizing agents, accelerators, activators, retarders (inhibitors), antidegradants, fillers, plasticizers, and special additives [10–13].

Two main types of additives are involved in the material's ability to dampen mechanical vibration. Firstly, the oil or plasticizer significantly contributes to determining the damping characteristics. The oil choice is usually decided by the required low temperature flexibility and by damping characteristics. Secondly, the carbon black type and its amount are significant factors that influence the damping characteristics of a rubber compound [14–16].

The aim of this study is to investigate mechanical and vibration damping properties of rubber composites containing different carbon black primary particle sizes with slightly different properties. In order to determine the influence of a carbon black structure on mechanical and vibration damping properties, the rubber mixture compositions were prepared with the same carbon black volume concentration.

2. Materials and Methods

2.1. Materials

Styrene butadiene rubber (SBR) SBR 1500 (Synthos Kralupy a.s., Kralupy nad Vltavou, Czech Republic), was used as the base for rubber compounds. Carbon black grades N 110, N 330, N 550, and N 990 from CS Cabot, s.r.o. (Valašské Meziříčí, Czech Republic), were used as fillers. The additives, including ZnO from SlovZink a.s. (Košeca, Slovakia), stearic acid from Setuza a.s. (Ústí nad Labem, Czech Republic), N-tert-butyl-benzothiazole sulfonamide (TBBS) from Duslo a.s. (Šaľa, Slovakia), and sulphur type Crystex OT33 from Eastman Chemical company (Kingsport, Tennessee, USA), were also compounded. The rubber recipe is given in Table 1.

Table 1. Rubber compound recipe. SBR, styrene butadiene rubber; TBBS, N-tert-butyl-benzothiazole sulfonamide.

Ingredients	Loading [phr]
SBR	100
Carbon black	50
ZnO	3
Stearic acid	1
TBBS	1
Sulfur	1.75

Iodine adsorption (IA) and dibutyl phthalate absorption (DBPA) methods were used in order to characterize the carbon black more precisely. While the surface area could be characterized by the first method, the second one was used for secondary structure characterization. The IA of the studied carbon black types was 145, 82, 43, and 10 g × kg^{-1} for N 110, N 330, N 550, and N 990, respectively. The secondary structure was more similar for N 110, N 330, and N 550 carbon black types (113, 102, and 121 mL × 100 g^{-1}, respectively). However, the last filler type (N 990) was characterized by the lowest secondary structure, with a DBPA of 35 mL × 100 g^{-1}. The primary carbon black particle sizes were measured using JEM-2100 transmission electron microscope (JEOL Ltd., Tokyo, Japan) and the average diameters of the studied carbon black were about 15 nm for N 110, 27 nm for N 330, 55 nm for N 550, and 280 nm for N 990. The optimal filler concentration in a polymer matrix is strongly influenced by the stiffening effect (particle size, shape, and structure) of the filler. With decreasing particle size, the stiffening effect increases. A filler value of 50 phr was chosen as the optimal concentration for various types of carbon black to be able to evaluate their effects on the studied properties. At the given phr concentration, most of rubber mixtures exhibits optimal physical-mechanical properties.

2.2. Obtaining the Mixtures and Sample Preparation

The basic rubber compounds (SBR, carbon black, ZnO, and stearic acid) were mixed in a Banbury laboratory internal mixer (Farrel Pominy, Castellanza, Italy) (volume 0.41 L, fill factor 0.72, chamber temperature 70 °C) for 8 min at 80 rpm. Consequently, the compounds were cooled down to 80 °C over a period of 1 min on a double roll mill (150 mm × 330 mm, Farrel) at a rotation speed ratio of 1:1.2. Finally, the accelerator and sulphur were added and mixed for the next 6 min.

2.3. Methods

2.3.1. Curing Characterization

Curing characteristics were measured on the Premier MDR moving die rheometer from Monsanto Company (Dayton, Ohio, USA) at a constant temperature of 160 °C. Scorch (t_{s1}) and optimum cure (t_{90}) times were evaluated by these measurements.

Consequently, compression molding in a hydraulic press at 20 MPa and 160 °C was used for the test sheet preparation. The samples sheets (150 mm × 150 mm) for mechanical tests measured 1 and 2 mm in thickness, cylinders measured 10, 15, 20, and 80 mm in diameter for vibration tests, and cylinders measured 20 and 100 mm in diameter for cyclic pressure tests. The samples measuring up to 2 mm in thickness were molded over the given time period, which is given by the time t_{90}. The curing time for samples thicker than 2 mm was calculated as the value of time t_{90} with the addition of 1 min per each 1 mm of sample thickness.

2.3.2. Quasistatic Mechanical Properties

Tensile tests were performed on the T10D tensile testing machine from Alpha Technology (Hudson, Ohio, USA), according to ISO 37 standard. Ten specimens (type 2, elongation speed of 500 mm/min) were tested in this case. The average values of the measured quantities and their standard deviations were subsequently evaluated.

Shore A hardness was measured according to ISO 7619 standard. Samples measuring 6 mm in thickness were tested and the parameters (median and arithmetic average) were obtained from nine measured values.

Rubber rebound resilience was measured according to ISO 4662 standard. The sample measuring 10 mm in thickness was examined and the average values of the resilience, including its standard deviations from 6 values, were subsequently determined. These tests were carried out at an ambient temperature of 22 °C.

2.3.3. Dynamical Mechanical Properties

Dynamical mechanical analysis (DMA) of the vulcanized rubber samples measuring t = 1 mm in thickness and w = 6 mm in width was realized on the Mettler Toledo DMA 1 equipment (Columbus, Ohio, USA) in tensile mode. Firstly, the temperature sweep was measured in a temperature range of −80 °C to +20 °C, with a temperature rate of 2 °C/min, at a constant deformation rate of 5 µm and a frequency of 20 Hz. Secondly, the frequency sweep was studied at a constant deformation rate of 10 µm and at 25 °C, with varying frequency ranging from 0 to 250 Hz. Finally, the strain sweep was studied at a constant frequency of 20 Hz and at 25 °C, with varying strain ranging from 0 to 200 µm. Three measurements were taken for each compound.

The dynamic shear properties were measured on the Premier RPA rubber process analyzer from Alpha Technology (Hudson, Ohio, USA). Rubber compounds were firstly cured for t_{90} at 160 °C. The rubber sample was subsequently cooled down to 60 °C (cooling rate of 30 °C/min), followed by temperature stabilization for 3 min. Finally, the sample was deformed from 0.1% to 200% at frequency $f = 1$ Hz. Again, three measurements were performed for each compound.

2.3.4. The Hysteresis Characteristics

The hysteresis characteristics of the rubber compounds were evaluated on the ZwickRoell tensile testing machine (ZwickRoell, Ulm, Germany) in compression mode. The samples measuring 100 mm in diameter and 20 mm in thickness were tested. Cyclic deformation of 10% at 50 mm/min was applied to each sample. The total number of 100 cycles was performed to reach a steady state of the hysteresis loop. The tests were carried out at 23 °C. Cycling loading tests of vibration damping were also performed with the same machine. The rubber samples were cyclically loaded for 0; 100,000; 250,000; 500,000; and 750,000 cycles at a constant force of 10 kN, excitation frequency $f = 20$ Hz, and ambient temperature of 23 °C.

2.3.5. Mechanical Vibration Damping Measurement

In general, vibration measurement methods are divided into contacting and contactless types [17–20]. The contactless vibration measurement methods can be performed using inductive, capacitive, optical (e.g., laser triangulation and laser interference), and ultrasonic sensors. They can be applied to measure deflections of rotary components, as well as higher vibration amplitudes and excitation frequencies compared to the contacting vibration measurements, which can be performed using piezoelectric and microelectro mechanical system (MEMS) accelerometers. The accelerometers are mounted directly on the vibrating component and are being used more widely due to the rapid development of electronic technology, accompanying secondary instrument, and low-noise cables, and their high insulation resistance and small capacitance.

The vibration damping properties of the investigated rubber composites can be examined by different methods, namely using the free vibration method and by using harmonically excited vibration [21,22]. The free vibration method evaluates a material´s ability to dampen mechanical vibrations based on the logarithmic decrement δ and the damping ratio ζ. For harmonically excited vibration, which can be achieved under harmonic force or under the harmonic motion of a base, the vibration damping properties are characterized by the frequency dependencies of amplitude ratios (e.g., amplification factor or displacement transmissibility). The method involving harmonically excited vibration based on the response of a damped system under the harmonic motion of the base (so-called kinematic excitation), which is relatively simple, was applied in order to investigate the vibration damping properties of the studied rubber composites.

A material´s ability to damp mechanical vibration can be characterized by the transfer damping function D (dB), which is expressed by the following equation [23]:

$$D = 20 \cdot \log \frac{v_{01}}{v_{02}} \qquad (1)$$

where v_{01} is the velocity amplitude on the input (i.e., excitation) side of the tested sample and v_{02} is the velocity amplitude on the output side of the tested sample. For harmonically excited vibration, it is also possible to express the transfer damping function as follows:

$$D = 20 \cdot \log \frac{y_{01}}{y_{02}} = 20 \cdot \log \frac{a_{01}}{a_{02}} \qquad (2)$$

where y_{01} (a_{01}) is the displacement (acceleration) amplitude on the input side of the tested sample and y_{02} (a_{02}) is the displacement (acceleration) amplitude on the output side of the tested sample. There are three different types of mechanical vibration depending on the transfer damping function value, namely damped ($D > 0$), undamped ($D = 0$), and resonance ($D < 0$) vibration.

The mechanical vibration damping testing of the tested rubber composite materials was performed using the forced oscillation method. The transfer damping function was experimentally measured using the BK 4810 mini-shaker (Brüel and Kjær, Nærum, Denmark) in combination with a BK 3560-B-030 signal pulse multi-analyzer (Brüel and Kjær, Nærum, Denmark) and a BK 2706 power amplifier (Brüel and Kjær, Nærum, Denmark) at the frequency range of 2–3200 Hz (see Figure 1). Sine waves were generated by the mini-shaker. The acceleration amplitudes on the input and output sides of the investigated specimens were recorded by the BK 4393 A_1 and A_2 piezoelectric accelerometers (Brüel and Kjær, Nærum, Denmark). The accelerometers have these parameters: the frequency ranges from 0.5 to 16,500 Hz; the temperature range from −74 to 250 °C; and the weight is 2.4 g [24]. Measurements of the transfer damping function were performed for different inertial masses m (i.e., for 0, 90 and 500 g), which were located on the upper side of the harmonically loaded investigated samples (see Figure 1). Moreover, vibration damping properties of the investigated rubber samples with ground plane dimensions of 60 mm × 60 mm were performed for three different thicknesses (i.e., for 10, 15, and 20 mm) of these materials. The view of the experimental setup used for the vibration damping testing is shown in Figure 2. Each measurement was repeated 10 times at an ambient temperature of 22 °C.

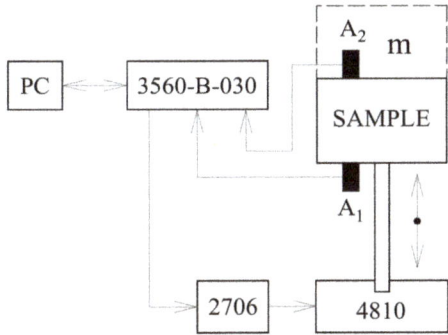

Figure 1. Schematic diagram of the measuring device.

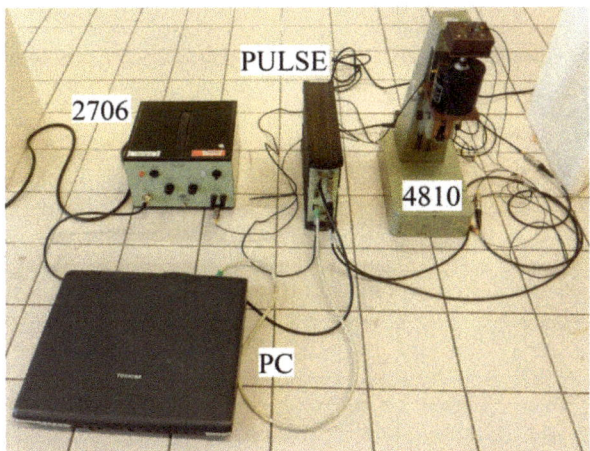

Figure 2. View of the experimental setup used for the vibration damping testing. Legend of the abbreviations: PC—personal computer; PULSE-signal pulse multi-analyzer; 2706—power amplifier; 4810—mini-shaker.

3. Results and Discussion

3.1. Curing Characteristics

The curing characteristics of the mixed compounds are given in Table 2. It is evident that the minimum torque M_L significantly decreased with an increase in the primary carbon black particle size. This is caused by the decreasing amount of the immobilized rubber chains on the decreasing carbon black surface, as well as the structure. The low aggregate structure and surface area of the N 990 type leads to weak interaction with rubber chains. Moreover, the thermal production process leads to a higher purity of the carbon black surface, with low content of active groups. This fact can result in a lower stiffening effect. However, in the case of the torque M_L, interactions between rubber chains and carbon black are mainly caused by physical forces, because the property is in an uncured state [25–28].

Table 2. Curing characteristics of the studied compounds.

Rubber Type	M_L [N·m]	M_H [N·m]	ΔM [N·m]	t_{s1} [min]	t_{90} [min]
N 110	0.27	1.71	1.44	2.6	10.5
N 330	0.24	1.45	1.21	2.6	11.5
N 550	0.23	1.51	1.28	3.0	12.8
N 990	0.12	1.07	0.95	3.3	12.6

The difference between the maximum (M_H) and the minimum (M_L) torques is marked as ΔM, which is a parameter demonstrating the degree of chemical crosslinking. The same amount and type of curing system was observed in the compounds; thus, the degree of ΔM is supposed to be comparable for all compounds. In reality, the ΔM is significantly different. Evidently, the reaction between the rubber and the curing system is one of the factors affecting the crosslinking process [28,29].

The other factor affecting the crosslinking process is the chemical reaction of the rubber with functional groups on the carbon black surface, which differs depending on the carbon black grade. It was found that these functional groups could have either positive or negative effects on the curing characteristics, depending on carbon black type [29]. Although the N 330 type has a higher surface area compared to the N 550 type, its curing level is lower due to the presence of cure-retarding groups [30]. This phenomenon causes various t_{90} values for the compounds in this study.

3.2. Quasi-Static Test Results

The parameters of mechanical properties of the SBR and carbon black compounds are presented in Table 3. The hardness testing is the most obvious mechanical test. In this case, the Shore A hardness was measured. The obtained hardness results decreased from 69 to 55 Shore A and corresponded to the carbon black particle size, ranging from the largest specific surface (N 110) to the smallest one (N 990). The higher specific surface led to a harder rubber compound. It is evident from the above results that the hardness generally decreased with a decrease in the carbon black surface area.

Table 3. Mechanical properties of the rubber compounds.

Rubber Type	Shore A [Sh A]	Stress at Break [MPa]	Strain at Break [%]	300% Modulus [MPa]	Resilience [%]
N 110	69 ± 1	23.0 ± 1.4	480 ± 36	13.0 ± 1.3	38 ± 1
N 330	66 ± 1	23.9 ± 0.9	520 ± 23	12.7 ± 1.0	39 ± 1
N 550	64 ± 1	19.8 ± 0.5	510 ± 25	11.9 ± 0.5	44 ± 1
N 990	55 ± 1	11.9 ± 2.7	640 ± 75	4.1 ± 0.4	51 ± 1

The tensile test was performed in order to characterize the basic mechanical properties of the rubber compounds. It should be noted that the compounds containing the carbon black with small primary particles (N 110 and N 330) showed the highest break stress (about 23 MPa), while the elongation was the lowest. The increasing primary carbon black particle size causes the decrease of the break stress. This is caused by the reinforcing ability of the carbon black particle size. The smaller primary particle sizes led to a higher specific surface area and to a stronger restriction of the elastomer chain mobility.

Here, the 300% modulus is the parameter characterizing the stiffness of rubber vulcanizates, representing the stress at 300% extension. In addition to previous parameters, it is influenced by the carbon black surface area. The highest value for this modulus was obtained for the N 110 type. The 300% modulus decreased with a decrease in the surface area as well.

The rebound resilience increased from 38% up to 51% for N 110 and N 990 carbon black types, respectively. Additionally, this property is influenced by the particle size, and thus, by the surface area of the carbon black. The stiffening effect of the N 110 type is stronger in comparison to the others. It provides harder rubber with lower rebound elasticity, because a larger part of the mechanical energy is transformed into heat.

Theoretically, the mechanical properties should show greater differences between the N 110 and N 330 carbon black types. Unfortunately, mechanical properties are strongly influenced by the level of filler dispersion in the matrix, which also depends on the carbon black particle size. With decreasing particle size, more energy is required to achieve high filler dispersion. Thus, a shorter mixing time leads to poorer filler dispersion. For this reason, the mechanical properties are reduced. On the other hand, a longer mixing time can cause polymer chain scission and also a decrease in the final properties. For this study, the rubber compounds were prepared under constant mixing conditions.

3.3. Dynamical Mechanical Properties

The effects of various carbon blacks in the SBR matrix on the temperature dependence of the elastic (storage) modulus is summarized in Figure 3. These curves describe the temperature region, where the hard and brittle material behavior is replaced by the rubbery or viscoelastic behavior. The modulus decreased strongly in the glassy transition region. The elastomer chain mobility, and thereby the glass temperature, is noticeably affected by various additives in the rubber compound. The presence of filler, the amount, an increase in the surface area, or the structure can restrict the chain's movement ability and lead to the shift of the glassy region to higher or lower temperatures [9].

During heating, the elastic modulus of the studied compounds decreased in the temperature range from −50 °C to −25 °C. The most rapid decrease of the elastic modulus was observed for the compound containing the N 990 carbon black type. On the other hand, the highest values of the elastic modulus were obtained for the compounds with the N 110 and N 330 carbon black types. This is connected with the decreasing carbon black particle sizes and the increasing surface area.

The glass transition temperature for the rubber compounds determined from the loss factor (*tan δ*) curve of three measurements was in the low temperature region, namely −31.9 ± 0.6 °C, −32.5 ± 0.5 °C, −32.2 ± 0.6 °C, and 31.1 ± 0.8 °C for the N 110, N 330, N 550, and N 990 types, respectively (see Figure 4). The differences in the glass temperature between compounds were less than 10%. The shift in glass transition temperature was probably caused by sample thickness inhomogeneity and the slightly different clamping forces of each compound. As a result, no significant effect was observed for the particle size on the glass transition temperature. In addition, the damping properties of the rubber compounds can be evaluated from this figure. It is evident that the loss factor increased with an increase in the primary particle size.

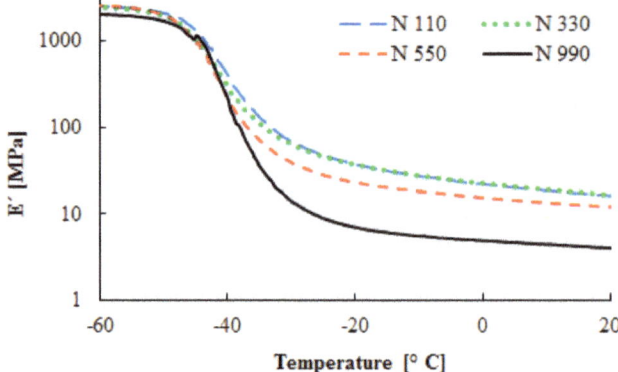

Figure 3. Temperature dependence of the tensile storage modulus.

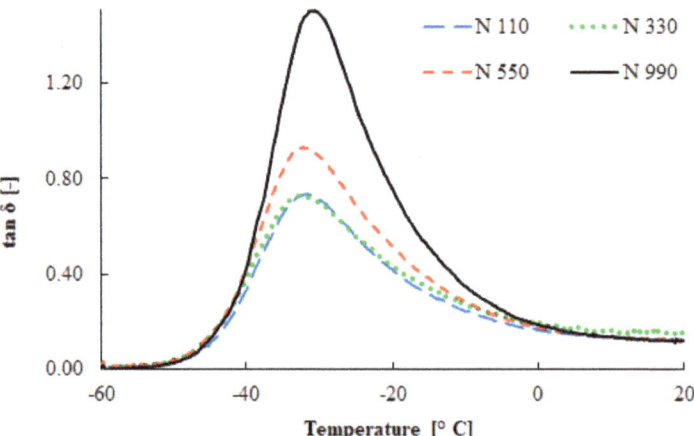

Figure 4. Temperature dependence of the loss factor.

The dynamic stiffness of the tested rubber mixtures is measured by the complex modulus of elasticity. The frequency dependencies of the storage modulus and the loss factor at 25 °C are demonstrated in Figures 5 and 6.

Significant changes of the storage modulus depending on the frequency for SBR compounds with the N 110, N 330, and N 550 carbon blacks are evident in Figure 5. There is a moderate increase of the storage modulus with the frequency up to 180 Hz. The filler with the smallest primary particle size (N 110) exhibited the highest storage modulus values, while N 990 gave the lowest values. This is connected with the filler–polymer interaction. This fact depends on the filler's primary particle size, as well as its structure [28].

It was found that the loss factor $\tan \delta$ at low frequencies up to 100 Hz showed similar behavior for all carbon black compounds (see Figure 6). Generally, the loss factor increased with an increase in the frequency. While for the more reinforced carbon black type (N 110) the increase of $\tan \delta$ was slower, in the case of the N 990 type the loss factor increased more significantly.

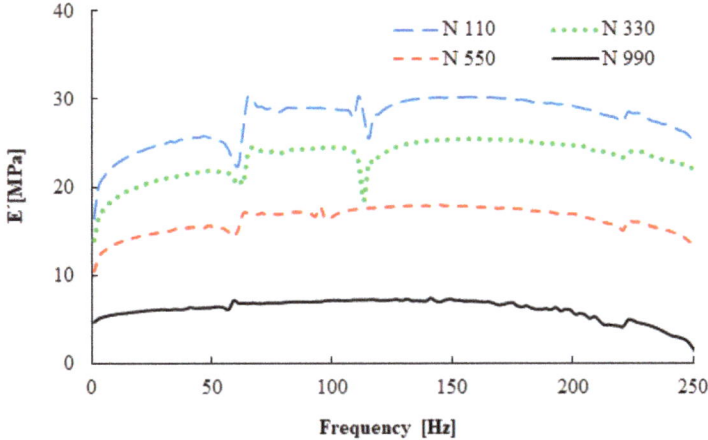

Figure 5. Frequency dependence of the tensile storage modulus (displacement x = 5 μm, T = 25 °C).

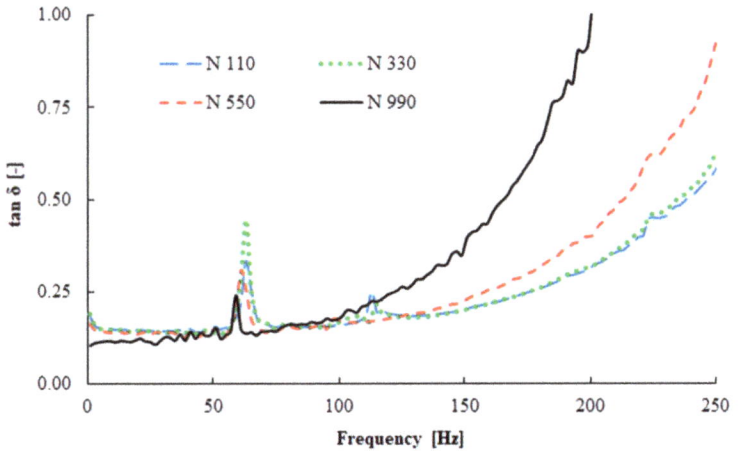

Figure 6. Frequency dependence of the loss factor (displacement x = 5 μm, T = 25 °C).

The next measurement was performed in order to evaluate the deformation dependence of the tested samples on their dynamic properties. The elastic modulus dependence on the displacement is shown in Figure 7. The effects of the carbon black particle size and structure were measured by this experiment. The compound with the smallest primary particle size gave the highest elastic modulus value, while that with the largest particle sizes showed almost no stiffening effect and had almost constant elastic modulus during the whole deformation range. High elastic moduli at low deformation levels are caused by the filler–filler interaction, as explained by Payne [31]. The smallest particles with higher carbon black structures created stronger filler–filler network in comparison with larger particles and a lower carbon black structures. With the increasing deformation, the filler network is destroyed and the interaction between the rubber and carbon black becomes the main factor.

Incorporated carbon black particles, which are already known as aggregates, create a filler–filler network. Inside the rubber compound, carbon black aggregates are formed by van der Waals forces into so called agglomerates. If a small deformation is applied, a high elastic modulus value is obtained.

The reason for this phenomenon is the strong interaction between filler particles that are not broken. From Payne's point of view, some of the rubber is immobilized on the filler surface, and in addition some of the rubber is also immobilized inside the branched structure of the agglomerate (known as occluded rubber). If the deformation increases, the agglomerates are broken into smaller sizes, and therefore the elastic modulus decreases. This phenomenon is caused by more mobile smaller units inside the rubber compound. At high deformation levels a plateau can be reached, whereby individual mobile aggregates units are set into motion. This behavior is known as the Payne effect [31–36].

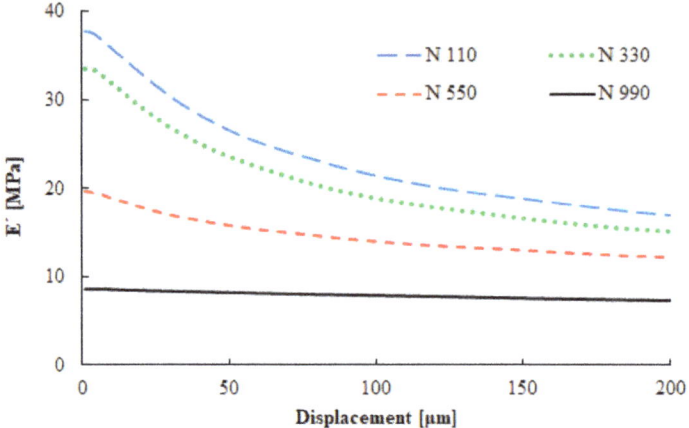

Figure 7. Displacement dependence of the tensile storage modulus (f = 20 Hz, T = 0 °C).

The dependence of the loss modulus on the sample deformation is presented in Figure 8. The loss modulus increased with a decrease in the primary particle size. The stronger filler–filler interaction led to higher energy dissipation. With the increasing deformation, the loss modulus maximum was found around the displacement amplitude of 35 μm for the compounds with the N 110 and N 330 carbon black types. The highest damping properties were obtained in this area. Apparently, the values of the loss modulus were low for the N 550 and N 990 carbon black types, while damping properties were especially negligible for the N 990 type.

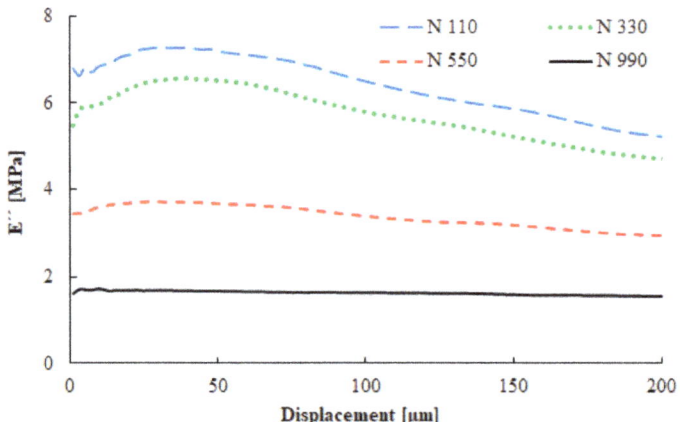

Figure 8. Displacement dependence of the tensile loss modulus (f = 20 Hz, T = 0 °C).

Shear amplitude deformation was measured using a rubber process analyzer. This test describes similar behavior of the rubber compounds as the DMA measurements, but in shear deformation mode.

The dependence of the filler–filler interaction for compounds with the N 110, N 330, N 550, and N 990 carbon black types is depicted in Figure 9. A strong dependence of the shear storage modulus on the filler particle size is visible, similarly to the DMA testing. Large carbon black particles (N 990) are characterized by a low reinforcing effect, while small particles (N 110) cause a pronounced increase of the storage modulus. Figure 10 shows the strain amplitude dependence of the loss modulus of four carbon black grades. Similarly to the DMA measurements, the shear loss modulus was highest for the N 110 type, while for the N 990 type it was the lowest. The maximum value of the loss modulus was observed at 1% strain, which is fully connected with the maximal damping properties.

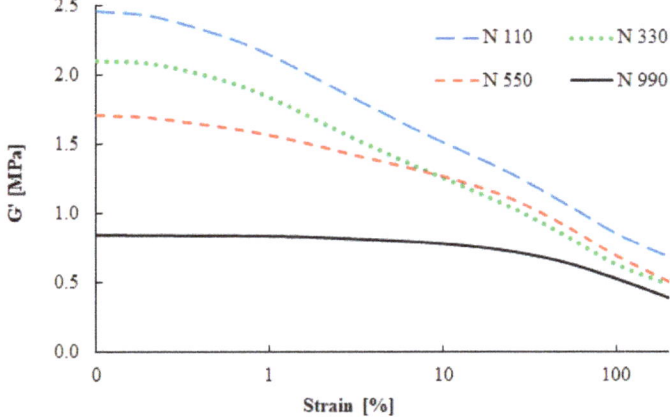

Figure 9. Strain amplitude dependence of the shear storage modulus.

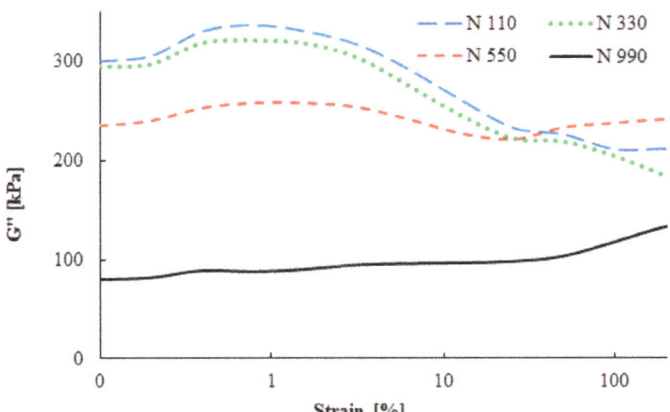

Figure 10. Strain amplitude dependence of the shear loss modulus.

3.4. The Hysteresis Characteristics

Due to the viscoelastic nature of the rubber vulcanizates, the stress–strain curves of the tested rubber create a hysteresis loop during the loading and unloading cycles. The hysteresis loop area corresponds to the energy dissipated into heat. The heat generation inside the rubber mixture can lead to it softening and even rupturing. The heat generation is affected by the polymer nature, curing level, and compound composition. This behavior is also known as the Mullins effect.

The hysteresis loops for the studied compounds are presented in Figure 11. The rubber was dynamically loaded in compression mode during 100 loading cycles. In this figure, the last cycle is recorded. Evidently, there are quite large differences among the carbon black types added to each compound. The largest area of the hysteresis loop was achieved for the N 110 carbon black type, which had the highest specific surface area, while the lowest heat generation was obtained for the N 990 type. An explanation of this phenomena was given by Fukahori [36]. According to this theory, rubber covers the carbon black surface in creation of the so-called bound rubber. This is an immobilized part of the rubber macromolecules that is physically connected with carbon black particles. During the loading of a polymer–carbon black structure, the orientation of this structure appears. If the unloading process is applied, the stress decreases faster compared to common macromolecular stress relaxation, and the decrease in the unloading curve is visible.

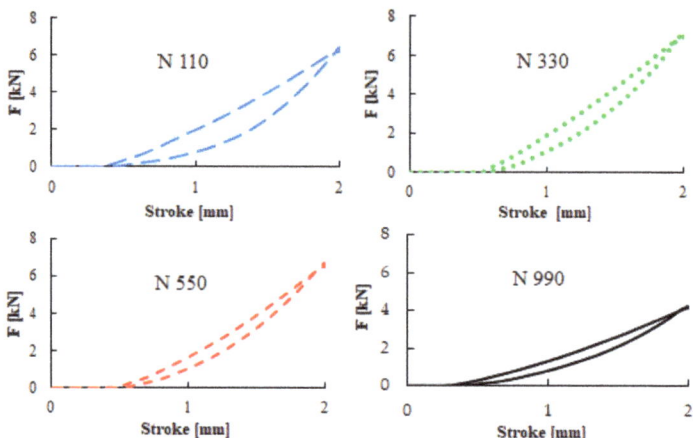

Figure 11. Hysteresis images of four carbon black types.

3.5. Vibration Damping Properties

Examples of the frequency dependencies of the transfer damping function of the tested rubber composites measuring $t = 10$ mm in thickness with different carbon black particle sizes are shown in Figure 12. It is evident from this comparison that the size of the carbon black particles has a significant influence on the vibration damping properties. It can be concluded that the material's ability to damp mechanical vibration generally increased with an increase in the carbon black particle size. This is caused by lower stiffness (or by higher damping) of the rubber composites, which were produced with larger carbon black particle sizes. These facts result in a higher transformation of input mechanical energy into heat during forced oscillations [37] and a decrease in the values of the damped and undamped natural frequencies [38]. Therefore, the first resonance frequency (f_{R1}) value was shifted to the left (see Figure 12) with an increase in the carbon black particle size, i.e., from 1520 (N 110) to 968 Hz (N 990), as indicated in Table 4. These findings are in excellent agreement with the results that were experimentally determined by the abovementioned methods, namely the hardness, tensile, shear, and viscoelastic measurements. It was verified in these cases that the increasing carbon black particle size led to a decrease of the break stress, the Shore A hardness, and the storage moduli E' and G'. In contrast, the rebound resilience was higher for these particle sizes.

Table 4. The first resonance frequency (f_{R1}) in Hz of the studied rubber composite materials, as induced by harmonic force vibration for different rubber thicknesses and inertial masses.

Rubber Type	Thickness [mm]	Inertial Mass [g]		
		0	90	500
N 110	10	1520 ± 41	1146 ± 18	430 ± 15
	15	1293 ± 29	1043 ± 18	419 ± 14
	20	1080 ± 15	928 ± 17	354 ± 12
N 330	10	1441 ± 32	1138 ± 19	412 ± 14
	15	1242 ± 22	957 ± 16	402 ± 13
	20	881 ± 16	520 ± 15	320 ± 9
N 550	10	1404 ± 27	900 ± 18	379 ± 13
	15	977 ± 18	737 ± 16	288 ± 10
	20	718 ± 16	456 ± 12	233 ± 8
N 990	10	968 ± 20	848 ± 16	298 ± 10
	15	802 ± 19	635 ± 15	258 ± 9
	20	693 ± 15	310 ± 11	231 ± 7

The vibration damping properties of the investigated harmonically loaded composite rubber samples are also influenced by their thickness t, the excitation frequency f, and the inertial mass m. The effect of the inertial mass on the vibration damping properties of sample N 330 is shown in Figure 13. It is visible that better damping properties were obtained with higher inertial mass m, which led to a decrease in the undamped natural frequency, and thus, the damped frequency. This is due to the fact that the natural frequency of an undamped system is proportional to the square root of the material stiffness for the applied inertial mass [22]. For this reason, the inertial mass has a positive influence on vibration damping, which is reflected by a shift of the first resonance frequency peak position to lower frequencies, i.e., by the decrease of the f_{R1} (see Table 4) from 1441 ($m = 0$ g) to 412 Hz ($m = 500$ g). The vibration isolation properties of the investigated rubber composites are also significantly influenced by their thickness t, as shown in Figure 14 for the N 330 sample with an inertial mass m of 90 g. It is evident that the higher material thickness led to lower values of the f_{R1}, i.e., from 1138 ($t = 10$ mm) to 520 Hz ($t = 20$ mm), as indicated in Table 4. For this reason, the rubber thickness generally has a positive influence on vibration damping properties. It is also visible from Figures 12–14 that the material´s ability to damp mechanical vibration is significantly influenced by the excitation frequency f. It is evident that resonant mechanical vibration ($D < 0$) was achieved at low excitation frequencies, depending on the rubber sample type, the thickness t, and the inertial mass m. For example, for the N 110 sample type measuring $t = 10$ mm in thickness and without inertial mass ($m = 0$ g), the resonant mechanical vibration was observed at frequencies $f < 2950$ Hz (see Figure 12). For the N 990 sample type measuring $t = 20$ mm in thickness and with inertial mass $m = 500$ g, the resonant mechanical vibration was achieved at considerably lower excitation frequencies (at $f < 330$ Hz). In contrast, damped mechanical vibration ($D > 0$) was generally achieved at higher excitation frequencies (see Figures 12–14).

The material´s ability to damp mechanical vibration was also evaluated for the tested material samples, which were harmonically loaded by the compression force with an amplitude of 10 kN at an excitation frequency of 20 Hz. In the case of the N 990 sample type, it was not possible to perform this evaluation due to the low stiffness of this rubber sample compared to the other tested rubber sample types. The frequency dependencies of the transfer damping function of the investigated rubber samples (thickness $t = 20$ mm, inertial mass $m = 90$ g) after 750,000 loading cycles are demonstrated in Figure 15. Again, as in the case of the cyclically unloaded rubber samples (see Figure 12), the material´s ability to dampen mechanical vibration increased with an increase in the carbon black particle size. For this reason the rubber composites, which were produced with larger carbon black particle sizes, exhibited lower stiffness, resulting in a decrease of the first resonance frequency peak position to lower excitation frequencies. As shown in Table 5, similar results were obtained independently of the number of loading cycles.

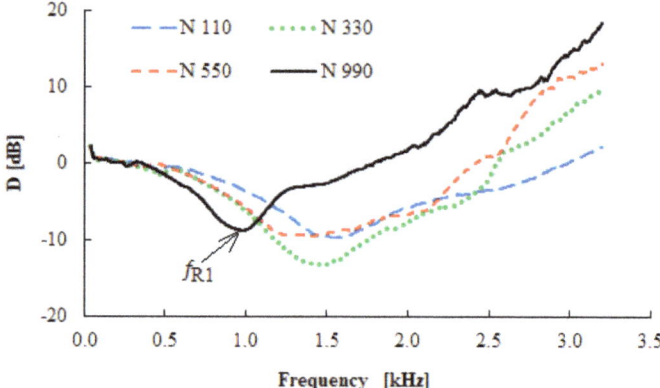

Figure 12. Frequency dependence of the transfer damping function (D) for the tested rubber composite measuring t = 10 mm in thickness, without inertial mass (m = 0 g).

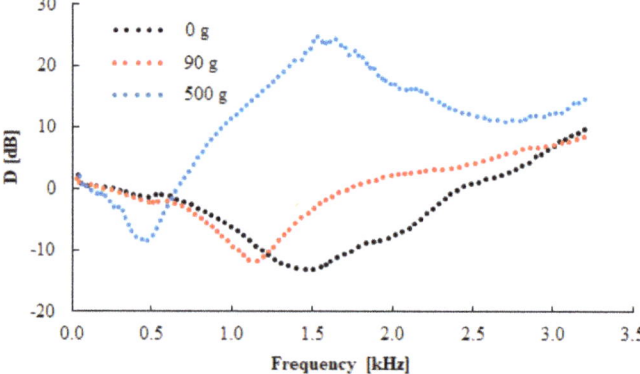

Figure 13. Frequency dependencies of the transfer damping function (D) for the tested N 330 rubber composite measuring t = 10 mm in thickness, loaded with different inertial masses.

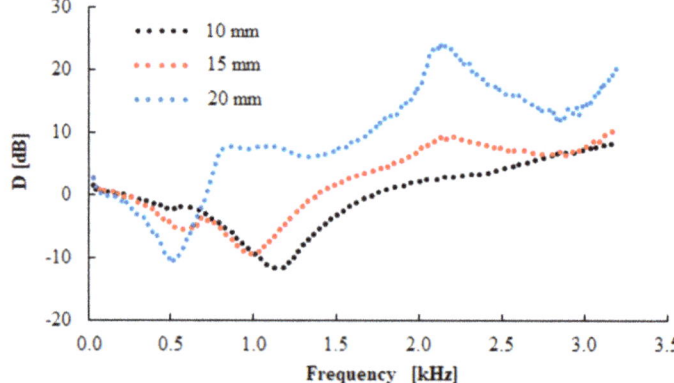

Figure 14. Frequency dependencies of the transfer damping function (D) for the tested N 330 rubber composites of different thicknesses and loaded with inertial mass m = 90 g.

Table 5. The first resonance frequency (f_{R1}) in Hz of the studied rubber composite materials measuring t = 20 mm in thickness, as induced by harmonic force vibration for different inertial masses and numbers of loading cycles.

Number of Cycles	Rubber Type	Inertial Mass [g]		
		0	90	500
0	N 110	1080 ± 15	928 ± 17	354 ± 12
	N 330	881 ± 16	520 ± 15	320 ± 9
	N 550	718 ± 16	456 ± 12	233 ± 8
100,000	N 110	1029 ± 17	563 ± 14	328 ± 11
	N 330	534 ± 15	507 ± 13	301 ± 10
	N 550	430 ± 13	378 ± 11	227 ± 9
250,000	N 110	482 ± 15	430 ± 12	245 ± 10
	N 330	457 ± 12	405 ± 11	241 ± 9
	N 550	381 ± 10	356 ± 10	192 ± 8
500,000	N 110	434 ± 11	388 ± 10	237 ± 8
	N 330	399 ± 11	371 ± 11	231 ± 8
	N 550	372 ± 10	237 ± 9	140 ± 6
750,000	N 110	404 ± 12	380 ± 11	208 ± 8
	N 330	374 ± 11	338 ± 10	158 ± 7
	N 550	360 ± 11	214 ± 8	126 ± 5

The effect of the number of loading cycles on the vibration damping properties of the N 330 sample type is shown in Figure 16. It is evident that the vibration damping ability of the sample generally increased with an increase in the number of loading cycles, which led to a decrease of the first resonance frequency peak position to lower frequency values (see Table 5) with the increasing number of loading cycles. Therefore, the higher number of loading cycles led to a reduction in rubber sample stiffness, which was accompanied by better damping properties in this rubber sample. As shown in Table 5, similar findings were observed for the other tested rubber composites.

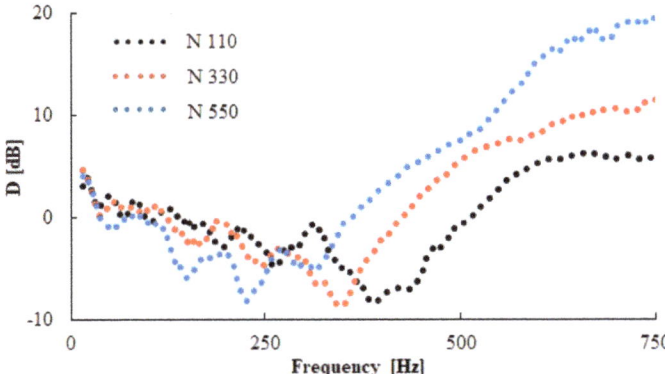

Figure 15. Frequency dependencies of the transfer damping function (D) for the tested rubber composites measuring t = 20 mm in thickness after 750,000 loading cycles, with inertial mass m = 90 g.

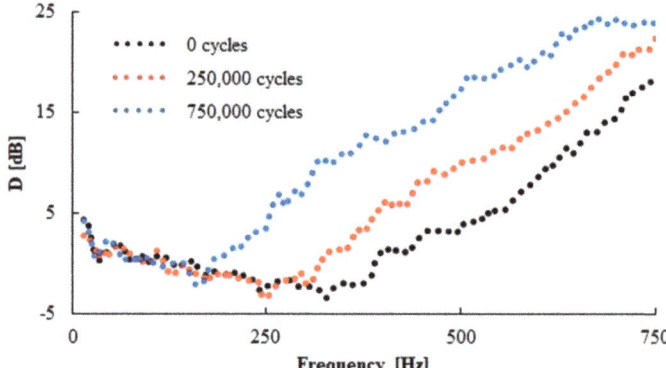

Figure 16. Effect of number of loading cycles on frequency dependencies of the transfer damping function (D) for the N 330 rubber composite measuring t = 20 mm in thickness, with inertial mass m = 500 g.

4. Conclusions

Mechanical vibrations are currently undesirable in many cases. Therefore, this vibration must be eliminated in appropriate ways. One of the possible elimination methods is the application of suitable vibro-insulating materials. This paper was focused on the study of the mechanical and vibro-isolation properties of rubber compounds containing different carbon black particle sizes. Nevertheless, their volume concentration was the same. On the basis of the evaluated measurements, it can be stated that the particle size of the carbon black in the rubber composites had a significant effect on the stiffness of the rubber, and thus on its mechanical and vibro-isolation properties.

The mechanical properties of the tested rubber mixtures were investigated for reflective elasticity, tensile, viscoelastic, and vibro-insulating properties. It was found in this work that the stiffness of the rubber samples generally decreased with an increase in the particle size, resulting in higher reflective elasticity and lower mechanical properties, including lower hardness (decrease from 69 to 55 Sh A), breaking stress, and real components of the complex modulus of elasticity (tensile and shear). On the other hand, the higher particle size in the rubber mixtures led to a greater loss factor and deformations in terms of sample breakage.

The above facts were in good agreement with the vibro-isolation tests for the observed rubber materials, which were examined using the forced oscillation method based on the transfer damping function. At the same time, a material's ability to dampen mechanical vibrations under dynamic stress is associated with the first resonance frequency, which is generally lower for materials that better dampen mechanical oscillation (or for materials with lower stiffness). Based on this method, it was verified that the first resonant frequency generally decreased with an increase in the carbon black particle size. Therefore, larger carbon black particles of the same volume concentration in rubber patterns contributed to better damping of mechanical vibrations, resulting in a higher transformation of input mechanical energy into heat under dynamic loading of these rubber composite samples. Vibration damping properties were also evaluated for the investigated rubber samples, which were harmonically loaded by a compression force. It can be concluded that a higher number of loading cycles led to a stiffness reduction in the investigated rubber composites, which was accompanied by a shift of the first resonance frequency peak position to lower excitation frequencies. Depending on the rubber type and the inertial mass, the decrease of the first resonance frequency after 750,000 loading cycles was between 37% and 65%. Furthermore, it has been found in this work that the material's ability to damp mechanical vibrations generally increased with an increase in the excitation frequency of mechanical vibration, inertial mass, and thickness of the investigated rubber samples.

Author Contributions: Conceptualization, M.P. and M.V.; methodology, M.P., M.V., and P.Z.; preparation of rubber composites, P.Z. and M.P.; experimentation, P.Z., M.P., M.V., and M.Ž.; data analysis, P.Z., M.P., and M.V.; resources, M.V. and M.P.; writing—original draft preparation, M.V., M.P., and D.M.; writing—review and editing, M.V., M.P., D.M., and P.Z. All authors have read and agreed to the published version of the manuscript.

Funding: This work was supported by the Research Center of Advanced Mechatronic Systems project, number CZ.02.1.01/0.0/0.0/16_019/0000867, within the Operational Program of Research, Development, and Education; and by grant IGA/FT/2018/008 and grant NPU I LO1504.

Conflicts of Interest: The authors declare that they have no conflicts of interest.

References

1. Stephen, N. On energy harvesting from ambient vibration. *J. Sound Vib.* **2006**, *293*, 409–425. [CrossRef]
2. Lapcik, L.; Cetkovsky, V.; Lapcikova, B.; Vasut, S. Materials for noise and vibration attenuation. *Chem. Listy* **2000**, *94*, 117–122.
3. Sakakibara, S. Properties of vibration with fractional derivative damping of order 1/2. *JSME Int. J. Ser. C Mech. Syst. Mach. Elem. Manuf.* **1997**, *40*, 393–399. [CrossRef]
4. Lee, K.H.; Kim, C.M. Application of viscoelastic damping for passive vibration control in automotive roof using equivalent properties. *Int. J. Automot. Technol.* **2005**, *6*, 607–613.
5. Elias, S. Seismic energy assessment of buildings with tuned vibration absorbers. *Shock. Vib.* **2018**, *2018*, 1–10. [CrossRef]
6. Qin, C.; Zhao, D.; Bai, X.; Zhang, X.; Zhang, B.; Jin, Z.; Niu, H. Vibration damping properties of gradient polyurethane/vinyl ester resin interpenetrating polymer network. *Mater. Chem. Phys.* **2006**, *97*, 517–524. [CrossRef]
7. Lee, H.J.; Lee, B.G.; Shin, D.Y. Vibration and impact noise damping properties of wood/polymer composites. *Mater. Sci. Forum* **2005**, *486*, 358–361. [CrossRef]
8. Goto, A.; Maekawa, Z.; Miyake, K. Analysis of vibration damping properties of hybrid composite with flexible matrix resin. *J. Soc. Mater. Sci. Jpn.* **1996**, *45*, 160–165. [CrossRef]
9. Perez, J.C.L.; Kwok, C.Y.; Senetakis, K. Effect of rubber content on the unstable behaviour of sand–rubber mixtures under static loading: a micro-mechanical study. *Géotechnique* **2017**, *68*, 1–14. [CrossRef]
10. Liu, Z.; Zhang, Y. Enhanced mechanical and thermal properties of SBR composites by introducing graphene oxide nanosheets decorated with silica particles. *Compos. Part A Appl. Sci. Manuf.* **2017**, *102*, 236–242. [CrossRef]
11. Berki, P.; Göbl, R.; Karger-Kocsis, J. Structure and properties of styrene-butadiene rubber (SBR) with pyrolytic and industrial carbon black. *Polym. Test.* **2017**, *61*, 404–415. [CrossRef]
12. Xu, H.; Han, J.; Fang, L.; Shen, F.; Wu, C. Effect of grafted carbon black on properties of vulcanized natural rubber. *Polym. Bull.* **2007**, *58*, 951–962. [CrossRef]
13. Nayak, S.; Chaki, T.K. Effect of hybrid fillers on the nonlinear viscoelasticity of rubber composites and nanocomposites. In *Non-Linear Viscoelasticity of Rubber Composites and Nanocomposites*; Springer International Publishing: Cham, Switzerland, 2014; pp. 135–160.
14. Omnès, B.; Thuillier, S.; Pilvin, P.; Grohens, Y.; Gillet, S. Effective properties of carbon black filled natural rubber: Experiments and modeling. *Compos. A Appl. Sci. Manuf.* **2008**, *39*, 1141–1149. [CrossRef]
15. Pan, X.-D.; Kelley, E.D. Enhanced apparent payne effect in a gum vulcanizate at low temperature approaching glass transition. *Polym. Eng. Sci.* **2003**, *43*, 1512–1521. [CrossRef]
16. Mandal, U.K.; Tripathy, D.K.; De, S.K. Effect of carbon-black fillers on dynamic-mechanical properties of ionic elastomer based on carboxylated nitrile rubber. *Plast. Rubber Compos. Process. Appl.* **1995**, *24*, 19–25.
17. Lénárt, J. Contactless vibration measurement using linear CCD sensor. In Proceedings of the 13th International Carpathian Control Conference (ICCC), High Tatras, Slovakia, 28–31 May 2012; pp. 426–429.
18. Cao, Y.; Rong, X.L.; Shao, S.J.; He, K.P. Present situation and prospects of vibration sensors. In Proceedings of the 2012 International Conference on Computer Distributed Control and Intelligent Environmental Monitoring, Zhangjiajie, Hunan, China, 5–6 March 2012; pp. 515–518.
19. Volpe, R.; Ivlev, R. A survey and experimental evaluation of proximity sensors for space robotics. In Proceedings of the 1994 IEEE International Conference on Robotics and Automation, San Diego, CA, USA, 8–13 May 1994; pp. 3466–3473.

20. Chaurasiya, H. Recent trends of measurement and development of vibration sensors. *Int. J. Com. Sci* **2012**, *9*, 1–6.
21. Botelho, E.C.; Campos, A.N.; de Barros, E.; Pardini, L.C.; Rezende, M.C. Damping behaviour of continuous fiber/metal composite materials by the free vibration method. *Compos. B Eng.* **2005**, *37*, 255–263. [CrossRef]
22. Rao, S.S. *Mechanical Vibrations*, 5th ed.; Pearson Education: Upper Saddle River, NJ, USA, 2011; Chapter 1.
23. Lapcik, L.; Vasina, M.; Lapcikova, B.; Valenta, T. Study of bread staling by means of vibro-acoustic, tensile and thermal analysis techniques. *J. Food Eng.* **2016**, *178*, 31–38. [CrossRef]
24. B&K Sound and Vibration Measurenet. Available online: https://www.bksv.com/en/products/transducers/vibration/Vibration-transducers/accelerometers/4393 (accessed on 12 May 2020).
25. Litvinov, V.M.; Steeman, P.A.M. EPDM–carbon black interactions and the reinforcement mechanisms, as studied by low-resolution ^{1}H NMR. *Macromolecules* **1999**, *32*, 8476–8490. [CrossRef]
26. Heinrich, G.; Vilgis, T.A. Contribution of entanglements to the mechanical properties of carbon black-filled polymer networks. *Macromolecules* **1993**, *26*, 1109–1119. [CrossRef]
27. Li, Z.H. Effects of carbon blacks with various structures on vulcanization and reinforcement of filled ethylene-propylene-diene rubber. *Express Polym. Lett.* **2008**, *2*, 695–704. [CrossRef]
28. Li, Q.; Ma, Y.; Wu, C.; Qian, S. Effect of carbon black nature on vulcanization and mechanical properties of rubber. *J. Macromol. Sci. B* **2008**, *47*, 837–846. [CrossRef]
29. Li, Z.; Zhang, J.; Chen, S. Effect of carbon blacks with various structures on electrical properties of filled ethylene-propylene-diene rubber. *J. Electrost.* **2009**, *67*, 73–75. [CrossRef]
30. Edwards, D.C. Polymer-filler interactions in rubber reinforcement. *J. Mater. Sci.* **1990**, *25*, 4175–4185. [CrossRef]
31. Lion, A.; Kardelky, C. The payne effect in finite viscoelasticity. *Int. J. Plast.* **2004**, *20*, 1313–1345. [CrossRef]
32. Markovic, G.; Marinovic-Cincovic, M.; Jovanovic, V.; Samarzija-Jovanovic, S.; Budinski-Simendic, J. Modeling of nonlinear viscoelastic behavior of filled rubbers. In *Non-Linear Viscoelasticity of Rubber Composites and Nanocomposites*; Springer International Publishing: Cham, Switzerland, 2014; pp. 193–271.
33. Drozdov, A.; Dorfmann, A. The payne effect for particle-reinforced elastomers. *Polym. Eng. Sci.* **2002**, *42*, 591–604. [CrossRef]
34. Wang, J.; Hamed, G.R.; Umetsu, K.; Roland, C. The payne effect in double network elastomers. *Rubber Chem. Technol.* **2005**, *78*, 76–83. [CrossRef]
35. Bezerra, F.D.O.; Nunes, R.C.R.; Gomes, A.S.; Oliveira, M.G.; Ito, E.N. Payne effect in NBR nanocomposites with organofilic montmorillonite. *Polimeros* **2013**, *23*, 223–228.
36. Fukahori, Y. New progress in the theory and model of carbon black reinforcement of elastomers. *J. Appl. Polym. Sci.* **2004**, *95*, 60–67. [CrossRef]
37. Júnior, J.H.A.; Júnior, H.L.O.; Amico, S.C.; Amado, F.D.R. Study of hybrid intralaminate curaua/glass composites. *Mater. Des.* **2012**, *42*, 111–117. [CrossRef]
38. Rajoria, H.; Jalili, N. Passive vibration damping enhancement using carbon nanotube-epoxy reinforced composites. *Compos. Sci. Technol.* **2005**, *65*, 2079–2093. [CrossRef]

© 2020 by the authors. Licensee MDPI, Basel, Switzerland. This article is an open access article distributed under the terms and conditions of the Creative Commons Attribution (CC BY) license (http://creativecommons.org/licenses/by/4.0/).

Article

Friction, Abrasion and Crack Growth Behavior of In-Situ and Ex-Situ Silica Filled Rubber Composites

Sankar Raman Vaikuntam [1,2], Eshwaran Subramani Bhagavatheswaran [1,2], Fei Xiang [1,2], Sven Wießner [1,2], Gert Heinrich [1,2], Amit Das [1,3] and Klaus Werner Stöckelhuber [1,*]

1. Leibniz-Institut für Polymerforschung Dresden e. V., Hohe Straße 6, 01069 Dresden, Germany; saram.tnv@gmail.com (S.R.V.); sb_eshwaran89@yahoo.co.in (E.S.B.); Xiang@ipfdd.de (F.X.); wiessner@ipfdd.de (S.W.); gheinrich@ipfdd.de (G.H.); das@ipfdd.de (A.D.)
2. Fakultät Maschinenwesen, Technische Universität Dresden, 01062 Dresden, Germany
3. Department of Materials ScienceTampere University, Korkeakoulunkatu 16, 33101 Tampere, Finland
* Correspondence: stoeckelhuber@ipfdd.de; Tel.: +49-351-4658-579

Received: 4 November 2019; Accepted: 23 December 2019; Published: 7 January 2020

Abstract: The article focuses on comparing the friction, abrasion, and crack growth behavior of two different kinds of silica-filled tire tread compounds loaded with (a) in-situ generated alkoxide silica and (b) commercial precipitated silica-filled compounds. The rubber matrix consists of solution styrene butadiene rubber polymers (SSBR). The in-situ generated particles are entirely different in filler morphology, i.e., in terms of size and physical structure, when compared to the precipitated silica. However, both types of the silicas were identified as amorphous in nature. Influence of filler morphology and surface modification of silica on the end performances of the rubbers like dynamic friction, abrasion index, and fatigue crack propagation were investigated. Compared to precipitated silica composites, in-situ derived silica composites offer better abrasion behavior and improved crack propagation with and without admixture of silane coupling agents. Silane modification, particle morphology, and crosslink density were identified as further vital parameters influencing the investigated rubber properties.

Keywords: elastomers; in-situ silica; friction; abrasion; tear fatigue test

1. Introduction

Rubber is one of the most important polymer materials playing an essential role in dynamic applications like tires, belts, diaphragms, and seals. Rubber products are continuously subjected to repeated cyclic deformation, according to the service conditions. In most areas, due to the working conditions like rubbing, abrading, chunking, and tearing the product undergoes mechanical failure very easily. Such failure mainly occurs due to continuous friction, wear, abrasion, and crack formation. Friction and wear often occurs when two surfaces contact each other [1]. Movements between the two contacting surfaces cause cutting or damage of the rubber surface, usually known as abrasion [2]. The continuous cutting of the soft rubber material generates small cracks, which propagates faster when the rubber product is subjected repeatedly to dynamic working conditions. The mechanical performance of rubber composites depends on several material parameters like type of elastomers, type of filler, filler content, filler size and structure, state of filler dispersion, interaction between polymer and filler, type of plasticizer, and the nature of crosslinks and its density [3]. Natural rubber is one of the main rubbers used in truck tire applications due to its excellent self-mechanical reinforcement and crack growth resistance, which are attributed to its strain induced crystallization behavior [4,5]. Incorporation of fillers like carbon black and precipitated silica into the natural rubber further improves the reinforcement characteristics. In recent years, novel nanofiller systems such as layered silicates, carbon nanotubes, or graphene nanoplatelets have been emerging, which, due to their nanoscale and their special aspect ratios, offer reinforcements beyond carbon black or silica [6–12]. Non-crystallizable rubbers like styrene butadiene rubber and acrylonitrile butadiene rubber have no self-reinforcement behavior. Therefore, the rubbers display poorer tear strength and crack propagation characteristics. These synthetic rubbers do not have the ability to crystallize during mechanical deformation [13]. Technically, for such non-crystallizable rubbers, the reinforcement is mainly enhanced only via a proper selection of fillers [14–16].

Carbon blacks are the major class of reinforcing materials for rubbers fulfilling the requisite properties for more than a century due to its ability to offer outstanding mechanical performance [17,18]. However, in the last few decades, precipitated silica is emerging as a potential candidate and has partially or fully replaced carbon black, mainly due to lower rolling resistance and improved wet grip properties in tire applications despite sensitively varying tear strength values and crack resistance properties [16,19]. However, poor nano/microscale dispersion and interaction with rubber is one of the critical issues encountered with silica and is strongly influencing the failure properties of rubber products. However, these issues have been overcome through multi-stage mixing and by incorporating suitable silane coupling agents. Scientifically, several assumptions are believed about the rubber product failure when rubbers are subjected to the mentioned severe working conditions. Some of them are: (1) molecular chains attempt to move in the stressed direction, causing chains slippage and reorientation, (2) crosslinks, entanglements, and filler particles hinder chain motion, which leads to a state of tension and local chain scission occurs, (3) scission of single polymer chain transfers the load to the neighboring chains, which propagates and the cumulative effect produces microvoids, and, lastly, (4) the microvoids grow to from microcracks, which is an irreversible process and these microcracks propagate and, ultimately, cause product failure [19].

For the current study, silica filled rubber composites are prepared by two different methods. First, commercial precipitated silica is directly incorporated into the rubber and next silica particles are generated inside the rubber by the sol-gel method [20]. More details concerning the preparation of sol-gel silica inside the rubber could be found in our earlier work [21]. The volume fraction of silica in both nanocomposites is the same and is confirmed through thermogravimetric analysis [21]. The particle morphology (size and structure), crosslink density, and state of silica dispersion of the prepared SSBR composites are entirely different. The synthesis of in-situ silica in the polymer matrix leads to trapping of polymer chains inside the silica aggregates as well as strong linkage of the polymer molecules at the silica surface. This can be considered as additional network nodes and is responsible for enhanced reinforcement of the rubber without a silane coupling agent. Usage of silane results in a further increase of the physical properties of the composites with in-situ silica due to improved silica-rubber interaction. Incorporation of a masterbatch containing in-situ silica particles leads to commercial precipitated silica to (a) enhanced silica dispersion, (b) possible higher silica loading in the elastomer matrix without impairing its processability, (c) the possibility to control and customize the particle size, and (d) energy-effective processing due to easy incorporation of the filler particles.

The main interest is to study how the filler size, structure, and crosslink density of the two different silica filled vulcanizates contribute to friction, abrasion, and crack propagation properties.

2. Materials and Methods

2.1. Materials

Solution styrene butadiene rubber (SSBR) BUNA 2525-2 VSL HM containing 25% vinyl content and 25% styrene content was supplied by Lanxess, Cologne, Germany. Tetrahydrofuran (THF), N-butyl amine, sulfur, zinc oxide, and stearic acid were acquired from Acros Organics, Germany. The vulcanizing accelerators N-cyclohexyl 2-benzothioazolesulfonamide (CBS-Vulkacit CZ) and diphenyl guanidine (DPG) were delivered by Lanxess, Germany. The precursor for the in-situ silica generation tetraethoxyorthosilicate (TEOS) was obtained from Sigma Aldrich (Munich, Germany) with a purity of 90%. The precipitated silica (Ultrasil VN-3) exhibiting a BET surface of 175 m^2/g and a purity of 99% and the silane coupling agent bis[3-(triethoxysilyl)propyl]tetrasulfide (TESPT) were kindly provided by Evonik Industries, Essen, Germany.

2.2. Preparation of Silica Rubber Composites

The preparation of the in-situ SSBR silica composites was performed in two separate steps. In the first step, in-situ silica-rubber masterbatches were produced and, in the second one, this polymer-silica masterbatch was mixed with the other rubber chemicals. For preparing masterbatches, 30 g of SSBR was dissolved in 300 mL of THF and, then, 0.1 mole of TEOS and 0.2 moles of water were added into a round flask. After forming a standard mixture, 0.025 moles of n-butylamine catalyst were added and the homogeneous solution was stirred at 60 °C for 4 h. The viscous liquid obtained was treated in an ultrasonic bath for 10 min to prevent any pre-agglomeration of silica particles inside the rubber. This solution was then carefully given into 900 mL of ethanol. This process immediately solidifies and precipitates the rubber mass. The precipitate was filtered and dried for two days at 40 °C in a circulating air oven. Using different amounts of TEOS and water (always at the molar ratio 1:2) leads to a controlled variability of volume fractions of silica in the rubber. Scheme 1 shows the chemical reaction for the formation of the silica particles and the particle growth [20].

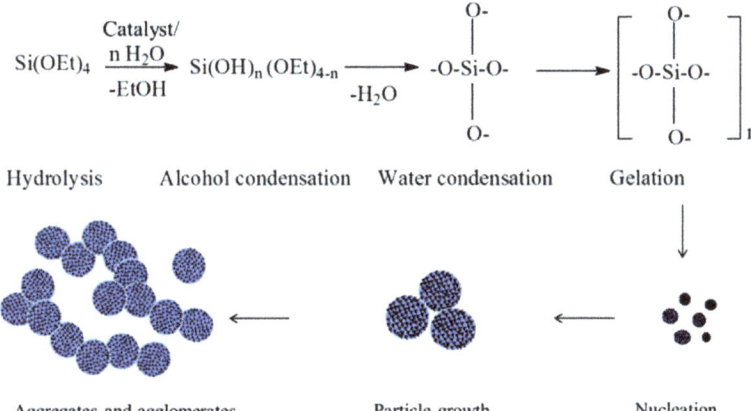

Scheme 1. Silica particle formation through sol–gel reaction reprinted from Reference [20] with permission from the Royal Society of Chemistry.

The second stage of preparation of in-situ silica composites consists of two sub-steps. In the first sub-step, in an internal mixer (HAAKE Rheomix 600P internal mixer, Karlsruhe, Germany), few grams of raw SSBR are added to the dried in-situ silica SSBR masterbatch. This is done to fine-tune the silica concentration in rubber and also to achieve the desired fill factor of 0.7 (70% filling). After adequate thorough blending, other rubber components as zinc oxide (2 phr), stearic acid (3 phr), and the TESPT coupling agent were added sequentially. This compounding process was carried out at 110 °C and at 80 rpm for 6 min. The mixture was dumped at a temperature of ~140 °C. In the second sub-step, the vulcanizing chemicals CBS (1.4 phr), DPG (1.7 phr), and sulfur (1.4 phr) were mixed into the compound in a two roll mill (Polymix-110L, Servitec, Wustermark, Germany) at 50 °C for 10 min with a constant friction of a 1:1.2 ratio.

Likewise, commercial precipitated silica composites were prepared by the incremental addition of zinc oxide, stearic acid, silica (Ultrasil-VN3), and silane TESPT into the SSBR rubber in the internal mixer. Mixing was carried out at 110 °C and 80 rpm for 10 min. For the addition of vulcanization chemicals in the two-roll mill, a process similar to that used for in-situ silica was applied. The finished compounds of in-situ and precipitated silica were then allowed to age for 24 h at room temperature. The stored rubber compounds were rheometrically analyzed in a rubber process analyzer (Scarabaeus

SIS-V50, Scarabaeus GmbH, Wetzlar, Germany) to determine their optimal vulcanization time TC_{90}. Vulcanization of these rubber samples was conducted at 160 °C into sheets of 2-mm thickness in a heated compression molding press (Fontijne TP1000, Delft, The Netherlands) with a force of 150 kN for optimum vulcanization time TC_{90} plus 1 min per mm thickness of the sample. The prepared silica-rubber samples are abbreviated as i-30, i-30s, x-30, x-30s. 'i' and 'x' represents in-situ and commercial precipitated silica filled composites, respectively. The number refers to the amount of silica in rubber (mentioned in 'phr'—parts per hundred rubber) and the suffix 's' after the number denotes rubber composites containing a TESPT silane coupling agent.

2.3. Characterizations of Silica-SSBR Nanocomposites

2.3.1. X-ray Diffraction (XRD)

X-ray diffraction analysis was carried out by means of a 2-circle diffractometer XRD 3003 T/T (Seifert-FPM, Freiberg, Germany) with Cu-Kα radiation at 30 mA and 40 kV from 2θ = 1° to 16° using 0.05° as the step length.

2.3.2. Dynamic Light Scattering Analysis (DLS)

The particle size and particle distribution of the silica powders was determined with a dynamic light scattering analyzer (Dyna Pro-nanostar, Wyatt instruments, Santa Barbara, CA, USA) from 0.2 to 2500 nm, using a laser wave length of 785 nm. Measurements are performed at 25 °C with dispersions of silica particles in ethanol. The DLS experiments were carried out for three samples. At least 25 scans per sample were performed.

2.3.3. Scanning Electron Microscopy (SEM)

Using an Ultra plus electron microscope from Carl Zeiss NTS GmbH, Oberkochen, Germany (3 kV, 30 μm aperture size, and SE2 detector), scanning electron microscope images of the silica-elastomer composites were recorded. Strip samples of these elastomer composites were broken after being exposed to liquid nitrogen. The respective fracture surface was then sputtered with 3-nm platinum with the BAL-TEC SCD 500 Sputter Coater and investigated under SEM at a zero angle.

2.3.4. Linear Friction Tester

Experiments on the friction properties of the SSBR-silica composites were carried out, using a tribometer-linear friction tester, located at the Deutsches Institut für Kautschuktechnologie (DIK), Hannover, Germany [22]. The instrument is shown schematically in Figure 1a. The test samples 20 mm × 20 mm × 10 mm by size were prepared in the compression molding machine (Fontijne TP1000, Delft, The Netherlands) at 160 °C at a force of 150 kN. The edges of the specimen are chamfered to avoid kinking of the elastomer specimen on contact with the asphalt surface. An example of the test specimen is shown in Figure 1b. Measuring temperatures from 2 °C to 100 °C, sliding velocities from 0.1 mm/s to 300 mm/s, and a variation of the load from 1 bar to 7 bar are possible at this experimental setup. Current experiments were performed at 15 °C with a load of 2 bars at velocities 0.1 mm/s to 300 mm/s. Measurements were performed on fine asphalt surface wetted with water to mimic wet friction properties of rubber composites similar like wet skid behavior of tires under ABS-braking conditions. It is commonly assumed that the ABS wet braking temperature for tire tread compound is ca. 40 °C (i.e., usually somewhat larger as the ambient temperature in mid-Europe) and the slipping velocity is around 1 m/s. It should be noted that, for the vehicle friction, the tread block slip velocity depends on time. In other papers, an effective (say time-averaged) slip velocity v = 0.3 m/s in water and v = 0.1 m/s in the dry state [23] was presumed. In reality, however, the tread blocks experience an inhomogeneous slip dynamic, where the slip velocity is near zero when a tread block enters the tire road footprint, while the slip velocity is typically in the order of a few m/s when the tread block is retracted from the footprint. The friction tester in our experiments is limited to velocities lower than 1 m/s. For this, we used the time-temperature superposition principle (Williams–Landel–Ferry (WLF) equation) to calculate the correct temperature for 10 mm/s (the measurable velocity of the machine). Therefore, the test temperature was determined using the universal WLF constants (C_1 = 17.44 and C_2 = 51.6 K) with help of the glass transition temperature of composites (T_g). The T_g values of the unfilled rubber as well as the silica-filled composites were determined by dynamic mechanical measurements (temperature sweep mode at a frequency of 10 Hz) [20].

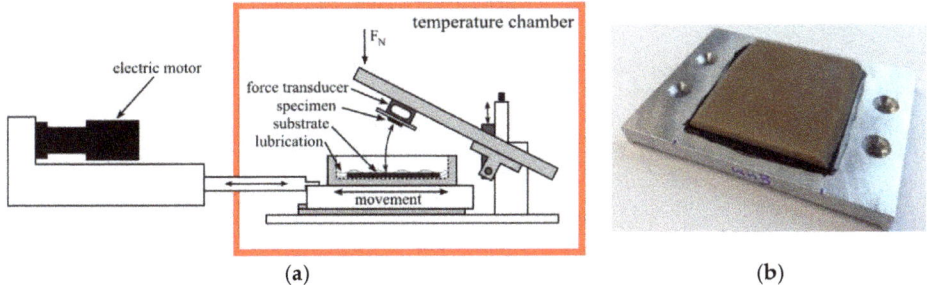

Figure 1. (a) Setup of the friction tester (schematically) reprinted from Reference [22] with permission from Elsevier and (b) an example test specimen.

Derived from the T_g of the elastomer compounds and the WLF principle, the temperature T_{calc} was calculated to be ca. 15 °C for a velocity of 10 mm/s (see Table 1). This temperature T_{calc} was further used in our experimental measurements.

Table 1. Glass transition temperatures of elastomer compounds and the calculated temperatures for the friction measurement by using the WLF principle.

Sample	T_g/°C	T_{calc} Calculated Temperature/°C	Hardness Shore A	Crosslink Density/ 10^{-4} g/cm^3
Gum	−27.8	14.96	42	1.463
i-30	−27.8	14.96	56	2.128
i-30s	−26.0	15.63	59	2.718
x-30	−28.2	14.81	56	1.571
x-30s	−26.8	15.33	57	2.202

2.3.5. Abrasion Test

The abrasion experiments were conducted in a rotary drum abrasion tester (Bareiss Prüfgerätebau GmbH, Oberdischingen, Germany, DIN ISO 4649 [24]) using 10 N static load. Circular samples with dimensions ⌀16.1 mm × 4.2 mm were tested. The abrasion resistance index (ARI) was calculated by the standard given formula in Equation (1).

$$ARI = \frac{\Delta m_r \times \rho_t}{\Delta m_t \times \rho_r} \times 100, \tag{1}$$

where Δm_r is the loss of mass from the standard rubber test specimen #1 in mg, ρ_r is the density of the standard elastomer sample #1 in g/cm^3, Δm_t is the mass loss of the test rubber piece in mg, and ρ_t is the density of the elastomer test sample in g/cm^3.

2.3.6. Tear Fatigue Analyzer

The fracture mechanics behavior of the rubber-silica composites was investigated using an Instron Electroplus E1000 (Darmstadt, Germany). Dumbbell-shaped pure shear specimens with length L_0 = 10 mm, thickness B = 1 mm, and width W = 80 mm (see Figure 2) were used to measure the crack propagation in these rubbers. The notch for crack initialization was made by hand for a length a_0 of 25 mm on one side of the pure shear sample, which generates a crack of the length a, and is expected to be adequately long related to the length (L_0) of the specimen. The crack propagation measurements were performed at room temperature (23 °C) using 1 N pre-force. Details of the experiments can be found in literature [25–28]. Rivlin and Thomas [29] stated that the tearing energy is the specific parameter of fatigue crack propagation in elastomers, which is defined as $T = -(\delta W/\delta A)_L$. Hereby, W is the elastic strain energy and A is the interfacial area of the propagated crack. Based on the assumption of Rivlin and Thomas [29], the pure shear specimen can be divided into four regions (see Figure 2). The elastomer in area A_1 is not strained. In region D, complicated deformation stress and strain fields occur due to the tip of the crack. In the C-region, pure shear deformation can be found and there is a region designated as A, where the edge effects occur. The deformation state in region D does not change but shift parallel in the crack growth direction, when the crack length increases by da. Then, the volume of the C-region is equal $L_0 \times B \times da$ and is, therefore, smaller and the stored elastic energy in this volume is released. Samples with pure shear geometry are used because, hereby, the tearing energy T_p does not depend on the crack length.

$$T_P = w \cdot L_0 \tag{2}$$

Hereby, w is the elastic pure shear strain-energy density, measured from the unloading cycle of the stress-strain curve of an un-notched sample. Two un-notched pure shear specimens with diverse widths ($W_1 = 80$ mm and $W_2 = 60$ mm) were employed for evaluating the pure shear elastic strain energy density w to eliminate the energy contribution of the edge effect region.

$$w = \frac{U_1 - U_2}{(W_1 - W_2) \cdot L_0 \cdot B} \tag{3}$$

where U_1 and U_2 are the elastic strain energies for the two different sample widths W_1 and W_2, respectively.

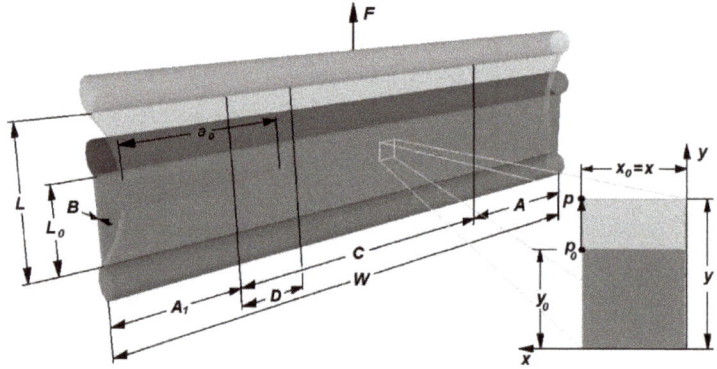

Figure 2. Schematic representation of a pure shear sample with depiction of the regions having different stress conditions (according to Reference [25]).

Paris and Erdogan [30] describe the correlation between a stable fatigue crack propagation rate da/dn and the tearing energy by a power law equation.

$$da/dn = \beta \cdot T^m \text{ for } T_1 \leq T \leq T_c, \tag{4}$$

β and m are, hereby, constants of the material. T_1 represents the beginning of the stable crack growth regime and T_c is the critical tearing energy, which characterizes the changeover to an unstable regime.

3. Results and Discussion

3.1. Morphology

Figure 3a displays the x-ray diffraction analysis (XRD) of in-situ and precipitated silica powder samples. The broad XRD band is located in the 2θ region of 15° to 35°. This confirms that the physical state of both silica types is entirely amorphous and there is no significant difference in the crystalline structure.

A dynamic light scattering measurement (DLS) was performed to characterize the two silica types in particle size (hydrodynamic diameter) and morphology. The precipitated silica powder particles (Ultrasil VN3) in Figure 3a display a bimodal distribution with a wide spectrum of different particle sizes. These silica particles are aggregated and agglomerated. The aggregate and agglomerate diameter sizes range from 150 to 1500 nm. The DLS results for the in-situ derived silica particles (see Figure 3b) give an average diameter of around 200 to 500 nm. The in-situ silica particles exhibit a mainly monodisperse or unimodal narrow particle distribution.

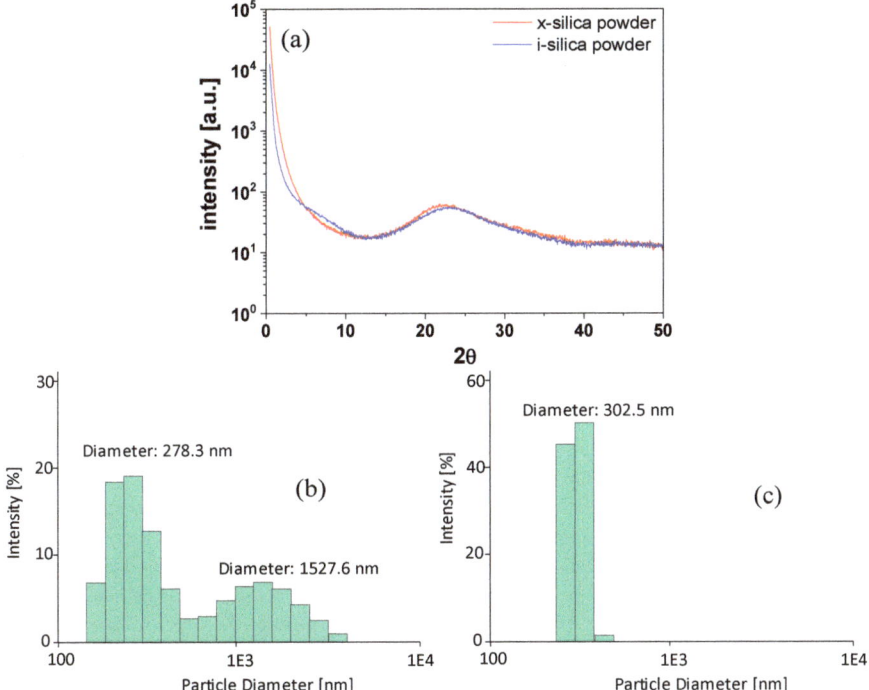

Figure 3. (a) X-ray diffraction patterns of in-situ silica and precipitated silica powders. Particle size histograms of (b) precipitated silica (Ultrasil-VN3) and (c) in-situ silica, from dynamic light scattering measurements.

SEM images of two silica powders and their respective SSBR rubber composites are given in Figure 4. Figure 4a shows how the precipitated silica particles look i.e., a network or chain-like structures with highly aggregated morphology. In Figure 4c, the in-situ silica particles display spherical aggregates with particle sizes from 200 to 400 nm. Nevertheless, the average primary particle sizes of both silica types appear to be in the order of ~10 to 15 nm. Figure 4b,d show the SEM pictograms of brittle fracture surfaces of 30 phr filled commercial precipitated silica and in-situ silica filled SSBR composites, respectively. The fractured images highlight the state of silica dispersion, which appears to be significantly different for either of the silica. An improved dispersion is found in the in-situ silica system.

Figure 4. SEM images of (**a**) pristine commercial precipitated silica powder, and (**b**) its respective SSBR composite. (**c**) In-situ silica powder (extracted from un-crosslinked rubber) and (**d**) its respective SSBR composite.

3.2. Friction Coefficients µ and Friction Curves

Figure 5 shows the experimental friction results measured at different sliding velocities with a normal load of 2 bar. The experimental results show that the unfilled elastomer exhibits a higher friction coefficient. An incorporation of silica in SSBR reduces the friction coefficient slightly. Precipitated silica without any silane exhibits a lower coefficient of friction in comparison to their respective silane-modified systems. The application of silane coupling agents to the system filled with precipitated silica has significantly improved the friction behavior in the high velocity range. The in-situ silica-filled systems exhibited lower friction coefficient values when compared to the precipitated silica system. It should be noted that incorporation of the silane coupling agent into the in-situ silica system did not affect the friction behavior. This result supports our earlier findings that in-situ silica has the capability to reinforce the rubber matrix without the need of any silane [20,21,31]. For good ABS (anti-lock braking or anti-skid braking) wet braking, the friction coefficient of the rubber system must be higher (or nearly equal to 1). Therefore, from the analysis, samples are rated as per their wet traction performance: Gum > x-30s > x-30 > i-30s > i-30.

On the other hand, rubber composites must display a lower friction coefficient to achieve lower rolling resistance properties. Lower rolling resistance contributes to better fuel efficiency of the vehicle. Lower rolling resistance can be observed when the rubber compounds display lower hysteresis and smaller tan δ in the temperature range around ca 60 °C as well as with a reduced rolling friction. From this perspective, the in-situ silica system will offer better low rolling resistance [21]. The samples rated as per their low rolling resistance (larger the better): Gum < x-30s < x-30 < i-30s < i-30.

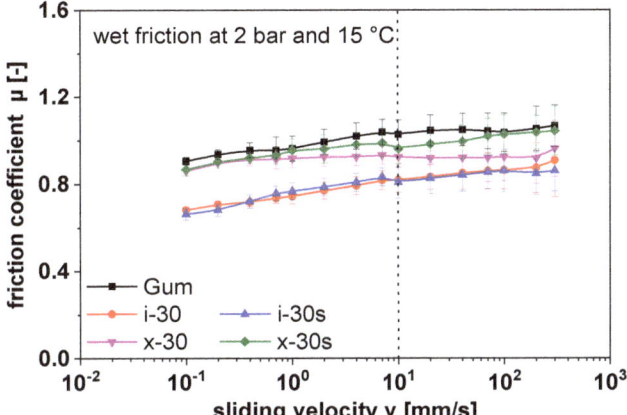

Figure 5. Friction behavior of unfilled rubber, in-situ silica, and precipitated silica with and without addition of a silane coupling agent. The dotted line corresponds to ABS wet braking conditions at 40 °C and a slipping velocity of 1 m/s.

3.3. Abrasion Test

Figure 6 shows the abrasion resistance index (*ARI*) of the unfilled vulcanizate, in-situ silica and precipitated silica-filled and their respective silane modified composites. The unfilled gum vulcanizate exhibits least *ARI* and this is due to the poor physical properties of the gum rubber, where the rubber surface is easily cut-out by the abrasion forces. Addition of silica as reinforcing filler into the SSBR-rubber improves the *ARI* values. In addition, increasing the amount of silica in rubber gradually increases the *ARI* values. However, the improvements in *ARI* are not significant without the presence of silanes. In the case of precipitated silica without silane, no improvements could be seen in the ARI values beyond the addition of 30 phr of silica. The incorporation of TESPT as a coupling agent in the silica-rubber system improves the abrasion properties further in both in-situ and precipitated silica composites. Experimental observations indicate improved abrasion properties for the precipitated silica compounds in comparison to them with in-situ silica. The effective particle size and surface area of the commercial precipitated silica are higher than in-situ silica, which could be the reason for the improved abrasion characteristics.

To understand the abrasion behavior in detail, the topology of abraded surfaces is investigated by SEM spectroscopy and their respective micrographs are given in Figure 7. The abraded surfaces of SSBR samples filled with precipitated silica and in-situ silica compounds without the coupling agent TESPT are found to be rough in comparison with the silane modified samples. Absence of silane as a coupling agent easily allows the rubber particles to cut-off due to the rubbing force. This is because of the high inhomogeneity in the elastomer material and the poor filler-rubber interaction. Silane modification of both silica composites greatly improves the surface wear mechanism. Formation of ridges or waviness protects the rubber surface from the further abrasion process by the gradual reduction of the surface contact area [32,33]. Such ridge topologies are formed on the surface of the samples when they are modified with a silane coupling agent. This also shows the improved interaction between silica and polymer assisted by the silane coupling.

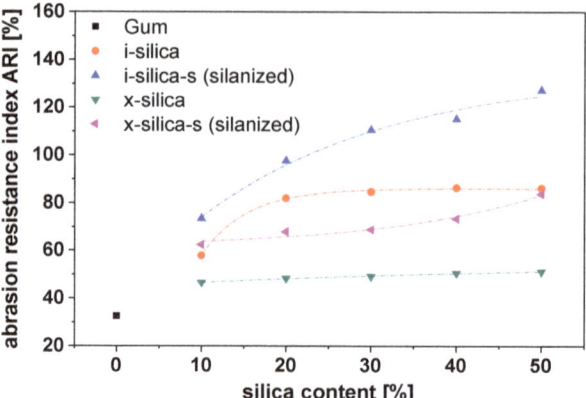

Figure 6. ARI of gum, in-situ, and precipitated silica compounds and its TESPT modified SSBR composites.

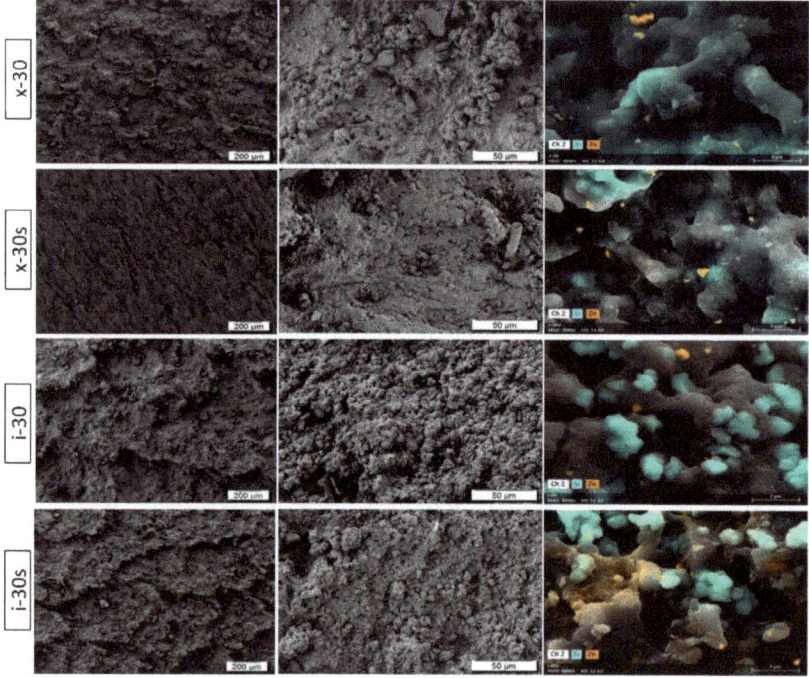

Figure 7. SEM and EDX images of abraded surfaces (after the DIN abrasion test) of 30 phr precipitated silica, in-situ silica, and their TESPT modified SSBR composites.

The lower ARI of the in-situ silica/SSBR composite could be attributed to the lower surface area and also due to the bigger particle size of the in-situ silica filler. In the EDX images in Figure 7, for samples without a silane coupling agent, it could be noticed that the silica filler particles are highly visible or exposed (noticeable as a blue color) after the abrasion experiment, since the silica particles are covered by a polymer layer. In in-situ silica-filled systems, the filler particles (noticeable as a blue color) are highly separated from the rubber matrix, when compared to the precipitated silica system.

This is an outcome of the poor abrasion behavior of the in-situ silica. In our previous investigation, we observed that the specific surface area of in-situ silica is very low in comparison to a precipitated silica system [34].

3.4. Tear Fatigue Analyzer (TFA)—Crack Propagation Measurements

A tear fatigue analysis was performed to understand the crack propagation behavior of different silica-filled SSBR composites and the influence of silica surface modification by a silane coupling agent. A logarithmic plot of the crack propagation rate (*da/dn*) in relation to the tearing energy (*T*) is known as a Paris plot and is given for all the samples in Figure 8. The rate of crack growth (*da/dn*) is directly proportional to the tearing energy *T* of the material. Therefore, all the samples are tested inside the stable crack growth regime. In Figure 8a, as expected, the unfilled gum sample shows a smaller tearing energy and a quicker crack formation compared to filled samples. It means that there is no active material to prevent the crack propagation in the gum vulcanizate. Incorporation of both in-situ and precipitated silica significantly improves the tear fatigue behavior. The main mechanism is the hindrance of filler particles that resist the crack tip growth [3]. Interestingly, the crack propagation behavior is found to be significantly different for the in-situ and precipitated silica, possibly due to the differences in their morphologies. The silane modification improved the crack propagation behavior for both the silica fillers. It is plausible that the significantly better crack resistance in silane-containing systems is a result of the establishment covalent chemical bonds between the elastomer matrix and filler particles by means of the bi-podal sulfur containing silane molecules. However, the in-situ silica system shows a higher crack growth rate than the precipitated silica system. The results, therefore, show that the state of the silica dispersion is not only a critical parameter for crack propagation, but that particle size, particle morphology, cross-linking density, and filler-polymer interfaces also play an important role.

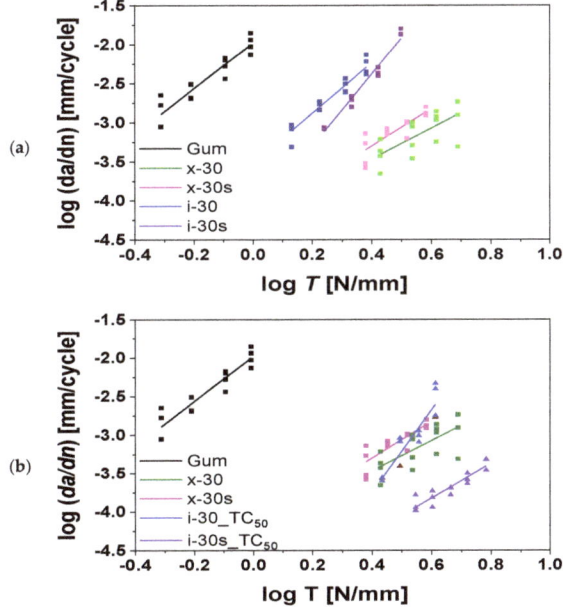

Figure 8. (**a**) Paris plots showing the crack propagation rate as a function of tearing energy for in-situ silica and precipitated silica-filled composites and (**b**) crack propagation of equally adjusted crosslink density of in-situ silica and precipitated silica compounds.

Our previous investigations of the SEM and TEM analysis distinctively states that particle size and morphology of precipitated and in-situ silica as well as the reinforcement characteristics are entirely different [20]. Moreover, the quantity of assessed cross-linking density is higher for the in-situ silica composites in comparison to precipitated silica composites. To eliminate the effect of different cross-linking density on the crack propagation behavior, the crosslink density of in-situ samples are reduced and matched with those of precipitated silica composites by modifying the state of cure from TC_{90} to TC_{50}. The amount of cross-linking density was measured by means of the equilibrium swelling method and the Flory-Rehner equation, which is described in detail in our previous paper [20]. The values of the cross-linking densities from this measurement are given in Table 2. The crack propagation behavior is completely different after adjusting the crosslink density of the in-situ silica composites. The plots in Figure 8b clearly indicate that the crack propagation behavior of in-situ silica composites is significantly improved. The in-situ silica compounds without silane modification yield better properties than the precipitated silica samples. However, the incorporation of silane in both of the silica systems reduces the crack propagation properties by a certain amount.

Table 2. Tear fatigue behavior of SSBR compounds filled with different silica systems with a reference to crosslinking density.

Sample	Cross-Linking Density /10^{-4} mol/cm^3	Slope
SSBR-Gum	1.463	2.9
x-30	1.571	3.2
x-30s	2.202	4.6
i-30	2.128	2.4
i-30s	2.718	1.9
i-30_TC_{50}	1.677	2.2
i-30s_TC_{50}	2.232	5.4

4. Conclusions

The present work has discussed the influence of in-situ silica and precipitated silica on friction, abrasion, and crack growth properties of SSBR rubber. For better understanding of the significance of silane in the silica-filled elastomer systems, the TESPT silane coupling agent was incorporated. The silanization affected the friction characteristics (wet grip performance) of precipitated silica-filled rubber. However, no significant effect was observed in the in-situ silica systems. Lower friction coefficients were observed for the in-situ silica systems, which is very vital for low rolling resistance (better fuel efficiency) applications. The precipitated silica compounds displayed better abrasion characteristics and the incorporation of a silane coupling agent plays a significant role by improving the filler-rubber interaction. The crack propagation results initially indicated that the precipitated silica system has higher resistance to crack propagation. However, upon comparing samples with similar crosslink densities, much better resistance to crack growth was observed for the in-situ silica systems, which is further improved by an additional silane modification. As a consequence of the results described in this paper, it can be stated that rubber compounds using a masterbatch containing in-situ silica (instead of the classical silica/silane mixing process) may have the potential to be used in the future in the industrial production of technical rubber articles or car tires. Yet, further research efforts are required for this implementation, in particular to ensure the scale-up of the process to industrial dimensions.

Author Contributions: Conceptualization, S.R.V., G.H., A.D., and K.W.S. Methodology, S.R.V., E.S.B., F.X., and S.W. Validation, G.H., A.D., and K.W.S. Formal Analysis, S.R.V., E.S.B., and F.X. Investigation, S.R.V. Resources, G.H. and S.W. Writing—Original Draft Preparation, S.R.V. Writing—Review and Editing, G.H., A.D., and K.W.S. Visualization, S.R.V. and E.S.B. Supervision, G.H., A.D., and K.W.S. Project Administration, S.R.V., A.D., and K.W.S. All authors have read and agreed to the published version of the manuscript.

Funding: This research received no external funding.

Acknowledgments: The authors express their thanks to Manfred Klüppel and Andrej Lang at Deutsches Institut für Kautschuktechnologie (DIK), Hannover for conducting the friction experiments and useful discussions. For the electron microscopy images, many thanks are given to Regine Boldt and Maria auf der Landwehr. The support of René Jurk in the Elastomer lab is gratefully acknowledged (all Leibniz-Institut für Polymerforschung Dresden).

Conflicts of Interest: The authors declare no conflict of interest.

References

1. Persson, B.N.J. On the theory of rubber friction. *Surf. Sci.* **1998**, *401*, 445–454. [CrossRef]
2. Schallamach, A. Friction and Abrasion of Rubber. *Rubber Chem. Technol.* **1958**, *31*, 982–1014. [CrossRef]
3. Persson, B.N.J.; Albohr, O.; Heinrich, G.; Ueba, H. Crack propagation in rubber-like materials. *J. Phys. Condens. Matter* **2005**, *17*, R1071. [CrossRef]
4. Rooj, S.; Das, A.; Morozov, I.A.; Stöckelhuber, K.W.; Stoček, R.; Heinrich, G. Influence of "expanded clay" on the microstructure and fatigue crack growth behavior of carbon black filled NR composites. *Compos. Sci. Technol.* **2013**, *76*, 61–68. [CrossRef]
5. Gent, A.N.; Lindley, P.B.; Thomas, A.G. Cut growth and fatigue of rubbers. I. The relationship between cut growth and fatigue. *J. Appl. Polym. Sci.* **1964**, *8*, 455–466. [CrossRef]
6. Galimberti, M. *Rubber—Clay Nanocomposites*; John Wiley & Sons: Hoboken, NJ, USA, 2011.
7. Bhowmick, A.K.; Bhattacharya, M.; Mitra, S.; Kumar, K.D.; Maji, P.K.; Choudhury, A.; George, J.J.; Basak, G.C. Morphology-property relationship in rubber-based nanocomposites—Some recent developments. In *Advanced Rubber Composites*; Heinrich, G., Ed.; Advances in Polymer Science; Springer: Berlin/Heidelberg, Germany, 2011; Volume 239.
8. Sadasivuni, K.K.; Ponnamma, D.; Thomas, S.; Grohens, Y. Evolution from graphite to graphene elastomer composites. *Prog. Polym. Sci.* **2014**, *39*, 749–780. [CrossRef]
9. Bokobza, L. Elastomers filled with carbon nanotubes. In *Polymer Nanotube Nanocomposites*; Mittal, V., Ed.; Wiley: Hoboken, NJ, USA, 2014; pp. 345–372.
10. Yaragalla, S.; Mishra, R.K.; Thomas, S.; Kalarikkal, N.; Maria, H. *Carbon-Based Nanofillers and Their Rubber Nanocomposites Fundamentals and Applications*; Elsevier: Amsterdam, The Netherlands, 2019.
11. Klüppel, M.; Möwes, M.M.; Lang, A.; Plagge, J.; Wunde, M.; Fleck, F.; Karl, C.W. Characterization and Application of Graphene Nanoplatelets in Elastomers. In *Designing of Elastomer Nanocomposites: From Theory to Applications*; Stöckelhuber, K., Das, A., Klüppel, M., Eds.; Advances in Polymer Science; Springer: Cham, Switzerland, 2016; Volume 275.
12. Najam, M.; Hussain, M.; Ali, Z.; Maafa, I.M.; Akhter, P.; Majeed, K.; Shehzad, N. Influence of silica materials on synthesis of elastomer nanocomposites: A review. *J. Elastom. Plast.* **2019**. [CrossRef]
13. Lake, G.J.; Lindley, P.B. Cut growth and fatigue of rubbers. II. Experiments on a noncrystallizing rubber. *J. Appl. Polym. Sci.* **1964**, *8*, 707–721. [CrossRef]
14. Rattanasom, N.; Saowapark, T.; Deeprasertkul, C. Reinforcement of natural rubber with silica/carbon black hybrid filler. *Polym. Test.* **2007**, *26*, 369–377. [CrossRef]
15. Dannenberg, E.M. The Effects of Surface Chemical Interactions on the Properties of Filler-Reinforced Rubbers. *Rubber Chem. Technol.* **1975**, *48*, 410–444. [CrossRef]
16. Reincke, K.; Grellmann, W.; Heinrich, G. Investigation of Mechanical and Fracture Mechanical Properties of Elastomers Filled with Precipitated Silica and Nanofillers Based upon Layered Silicates. *Rubber Chem. Technol.* **2004**, *77*, 662–677. [CrossRef]
17. Kraus, G. Reinforcement of Elastomers by Carbon Black. *Rubber Chem. Technol.* **1978**, *51*, 297–321. [CrossRef]
18. Medalia, A.I. Effect of Carbon Black on Dynamic Properties of Rubber Vulcanizates. *Rubber Chem. Technol.* **1978**, *51*, 437–523. [CrossRef]
19. Heinrich, G.; Horst, T.; Struve, J.; Gerber, G. Crack propagation in rubber-like materials. In Proceedings of the 11th International Conference on Fracture 2005 (ICF11), Turin, Italy, 20–25 March 2005; pp. 5497–5502, ISBN 978-1-61782-063-2.
20. Raman, V.S.; Das, A.; Stöckelhuber, K.W.; Eshwaran, S.B.; Chanda, J.; Malanin, M.; Reuter, U.; Leuteritz, A.; Boldt, R.; Wießner, S.; et al. Improvement of mechanical performance of solution styrene butadiene rubber by controlling the concentration and the size of in situ derived sol–gel silica particles. *RSC Adv.* **2016**, *6*, 33643–33655. [CrossRef]

21. Vaikuntam, S.R.; Bhagavatheswaran, E.S.; Stöckelhuber, K.W.; Wießner, S.; Heinrich, G.; Das, A. Development of high performance rubber composites from alkoxide-based silica and solution styrene-butadiene rubber. *Rubber Chem. Technol.* **2017**, *90*, 467–486. [CrossRef]
22. Lang, A.; Klüppel, M. Influences of temperature and load on the dry friction behaviour of tire tread compounds in contact with rough granite. *Wear* **2017**, *380–381*, 15–25. [CrossRef]
23. Tolpekina, T.V.; Persson, B.N.J. Adhesion and Friction for Three Tire Tread Compounds. *Lubricants* **2019**, *7*, 20. [CrossRef]
24. DIN ISO 4649:2002. *Rubber, Vulcanized or Thermoplastic—Determination of Abrasion Resistance Using a Rotating Cylindrical Drum Device*; ISO: Geneva, Switzerland, 2002.
25. Stoček, R.; Heinrich, G.; Gehde, M. The influence of the test properties on dynamic crack propagation in filled rubbers by simultaneous tensile and pure shear test mode testing. In *Const. Models Rubber VI*; Heinrich, G., Kaliske, M., Lion, A., Reese, S., Eds.; CRC Press: Boca Raton, FL, USA, 2010.
26. Stoček, R.; Heinrich, G.; Reincke, K.; Grellmann, W.; Gehde, M. Einfluss der Kerbeinbringung auf die Rissausbreitung in elastomeren Werkstoffen. *KGK Kautsch. Gummi Kunstst.* **2010**, *63*, 364–370.
27. Stoček, R.; Heinrich, G.; Reincke, K.; Grellmann, W.; Gehde, M. Rissausbreitung in Elastomeren Werkstoffen unter dynamischer Beanspruchung: Einfluss der Kerbeinbringung. *KGK Kautsch. Gummi Kunstst.* **2011**, *64*, 22–26.
28. Ghosh, P.; Stoček, R.; Gehde, M.; Mukhopadhyay, R.; Krishnakumar, R. Investigation of fatigue crack growth characteristics of NR/BR blend based tyre tread compounds. *Int. J. Fract.* **2014**, *188*, 9–21. [CrossRef]
29. Rivlin, R.; Thomas, A.G. Rupture of rubber. I. Characteristic energy for tearing. *J. Polym. Sci.* **1953**, *10*, 291–318. [CrossRef]
30. Paris, P.; Erdogan, F. A critical analysis of crack propagation laws. *J. Basic Eng.* **1963**, *85*, 528–533. [CrossRef]
31. Raman, V.S.; Rooj, S.; Das, A.; Stöckelhuber, K.W.; Simon, F.; Nando, G.B.; Heinrich, G. Reinforcement of Solution Styrene Butadiene Rubber by Silane Functionalized Halloysite Nanotubes. *J. Macromol. Sci. Part A* **2013**, *50*, 1091–1106. [CrossRef]
32. Gent, A.N.; Pulford, C.T.R. Mechanisms of rubber abrasion. *J. Appl. Polym. Sci.* **1983**, *28*, 943–960. [CrossRef]
33. Tabsan, N.; Wirasate, S.; Suchiva, K. Abrasion behavior of layered silicate reinforced natural rubber. *Wear* **2010**, *269*, 394–404. [CrossRef]
34. Vaikuntam, S.R.; Stöckelhuber, K.W.; Subramani Bhagavatheswaran, E.; Wießner, S.; Scheler, U.; Saalwächter, K.; Formanek, P.; Heinrich, G.; Das, A. Entrapped Styrene Butadiene Polymer Chains by Sol-Gel-Derived Silica Nanoparticles with Hierarchical Raspberry Structures. *J. Phys. Chem. B* **2018**, *122*, 2010–2022. [CrossRef]

© 2020 by the authors. Licensee MDPI, Basel, Switzerland. This article is an open access article distributed under the terms and conditions of the Creative Commons Attribution (CC BY) license (http://creativecommons.org/licenses/by/4.0/).

Article

Enhancement of Viscoelastic and Electrical Properties of Magnetorheological Elastomers with Nanosized Ni-Mg Cobalt-Ferrites as Fillers

Siti Aishah Abdul Aziz [1], Saiful Amri Mazlan [1,2,*], U Ubaidillah [3,4,*], Muhammad Kashfi Shabdin [2], Nurul Azhani Yunus [5], Nur Azmah Nordin [1], Seung-Bok Choi [6,*] and Rizuan Mohd Rosnan [7]

1. Engineering Materials and Structures (eMast) iKohza, Malaysia-Japan International Institute of Technology (MJIIT), Universiti Teknologi Malaysia, Jalan Sultan Yahya Petra, Kuala Lumpur 54100, Malaysia; aishah118@gmail.com (S.A.A.A.); nurazmah.nordin@utm.my (N.A.N.)
2. Advanced Vehicle System (AVS) Research Group, Malaysia – Japan International Institute of Technology (MJIIT), Universiti Teknologi Malaysia (UTM), Kuala Lumpur 54100, Malaysia; dekashaf@gmail.com
3. Department of Mechanical Engineering, Faculty of Engineering, Universitas Sebelas Maret, Surakarta 57126, Indonesia
4. National Center for Sustainable Transportation Technology (NCSTT), Bandung 40132, Indonesia
5. Department of Mechanical Engineering, Universiti Teknologi Petronas, Persiaran UTP, Seri Iskandar 32610, Malaysia; azhani.yunus@utp.edu.my
6. Department of Mechanical Engineering, Inha University, 253, Yonghyun-dong, Namgu, Incheon 22212, Korea
7. Jeol (Malaysia) Sdn. Bhd., 508, Block A, Kelana Business Center 97, Jalan SS7/2, Kelana Jaya, Petaling Jaya 47301, Malaysia; rizuanmr@jeolmal.com
* Correspondence: amri.kl@utm.my (S.A.M.); ubaidillah_ft@staff.uns.ac.id (U.U.); seungbok@inha.ac.kr (S.-B.C.)

Received: 8 August 2019; Accepted: 17 September 2019; Published: 28 October 2019

Abstract: Carbon-based particles, such as graphite and graphene, have been widely used as a filler in magnetorheological elastomer (MRE) fabrication in order to obtain electrical properties of the material. However, these kinds of fillers normally require a very high concentration of particles to enhance the conductivity property. Therefore, in this study, the nanosized Ni-Mg cobalt ferrite is introduced as a filler to soften MRE and, at the same time, improve magnetic, rheological, and conductivity properties. Three types of MRE samples without and with different compositions of Mg, namely $Co_{0.5}Ni_{0.2}Mg_{0.3}Fe_2O_4$ (A1) and $Co_{0.5}Ni_{0.1}Mg_{0.4}Fe_2O_4$ (A2), are fabricated. The characterization related to the micrograph, magnetic, and rheological properties of the MRE samples are analyzed using scanning electron microscopy (SEM), vibrating sample magnetometer (VSM), and the rheometer. Meanwhile, the effect of the nanosized Ni-Mg cobalt ferrites on the electrical resistance property is investigated and compared with the different Mg compositions. It is shown that the storage modulus of the MRE sample with the nanosized Ni-Mg cobalt ferrites is 43% higher than that of the MRE sample without the nanomaterials. In addition, it is demonstrated that MREs with the nanosized Ni-Mg cobalt ferrites exhibit relatively low electrical resistance at the on-state as compared to the off-state condition, because MRE with a higher Mg composition shows lower electrical resistance when higher current flow occurs through the materials. This salient property of the proposed MRE can be effectively and potentially used as an actuator to control the viscoelastic property of the magnetic field or sensors to measure the strain of the flexible structures by the electrical resistance signal.

Keywords: magnetorheological elastomer; filler; hetero-aggregation; nanosized Ni-Mg cobalt ferrites; electrical resistance; resistivity; rheological properties

1. Introduction

Magnetorheological elastomer (MRE) is one of the smart materials consisting of micron size magnetic particles embedded in rubber matrix. The rheological properties of MRE can be tuned continuously and reversibly by the external magnetic field. This kind of smart material has potential applications including dampers, sensors, and actuators for vibration systems [1–4]. Currently, researchers in the field of MRE are mainly focused on the effect of different elastomer matrices [5–7], types of magnetic particles [8–10], the effect of the particle sizes [11], and the shapes of the magnetic particles [12]. Silicon rubber (SR) has been widely used in MRE fabrication as the matrix due to its softness which can lead to a high MR effect [13–17]. Meanwhile, carbonyl iron particles (CIPs) have been widely used in MRE fabrication due to their high magnetic saturation, high permeability, and low remanence. In addition, additives and fillers also have been introduced in MRE fabrication to strengthen the interaction between the matrix and the CIP and hence improve mechanical, rheological, and electrical properties. On the other hand, a few researchers endeavored to fabricate MRE using the nanoparticles as magnetic particles to achieve large surface area, stable chemical reaction, and easy surface modifications [18–20]. Mordina et al. [21] introduced two types of different shapes of $CoFe_2O_4$, spherical and fiber in silicon-based nanocomposites, and found that the nanofibers enhanced the magnetic saturation as low as 5 wt.%, but the saturation tended to decrease at a higher loading of 10 wt.% due to the low formation of interconnected network of nanofibers. Despite this, by implementing the nanoparticles as an additive or filler in MRE, the magnetic and rheological properties of MRE can be appropriately tuned. Commercially available additives such as carbon black [22,23], graphite [4,24], ester [25,26], oils [27], and carbon nanotubes [28–30] have been used in MRE fabrication and have proven to enhance various performances of MRE as compared to the original MRE without these additives. Aziz et al. [31] introduced a few types of multiwall carbon nanotube (MWCNTs) as an additive in MRE fabrication. The results demonstrated that with the addition of 0.1 wt.% of MWCNTs, the magnetic saturation of MRE is slightly changed, as the saturation of MRE is increased up to 4%. In a similar work, Poojary et al. [28] fabricated MRE with different percentages of carbon nanotube (CNT), 0.1 wt.% to 0.5 wt.%, as additives and found that the stiffness and loss factor was the highest with the addition of 0.25 wt.% CNT, whereas MRE without CNT exhibited the highest MR effect of 48%. Padalka et al. [32] synthesized Fe, Co, and Ni nanowires and fabricated 10 wt.% of each in MRE composite. Their study focused on the strain and frequency with compression mode showing that MRE with Fe and Co exhibited the highest MR effect at 1% strain and tended to drop at 2% strain. Meanwhile, Yu et al. [33] used the nanoparticles for MRE fabrication as a coating material for the damping applications. They synthesized CIPs which were coated with Fe nanoflakes and demonstrated that the MR effect and damping properties increased up to 162% and 65%, respectively, with the incorporation of 6% CIP-Nano-Fe.

Generally, neat MRE is considered to be an insulator, and thus electrical property can be changed by adding the proper filling material for the development of long-lasting products for sensor applications. Thus far, a few researchers have attempted to synthesize MREs associated with the nanoparticles of additives or fillers to enhance various properties such as magnetic, electrical, optical, and catalytic properties, which are known to be useful in science and technological applications [21–23]. For instance, cobalt ferrites nanoparticles have been extensively studied due to their variety of applications including electrical and magnetic applications. Generally, the properties of cobalt ferrite can be altered by adding dopants and metal ions, adopting a variety of synthesis methods, and varying calcination temperature and other reaction conditions. Hossain et al. [22] synthesized various Ni substituted $Zn_{1-x}Ni_xFe_2O_4$ by solid-state reaction technique from stoichiometric amounts of Fe_2O_3, ZnO, and NiO of 99.99% purity and showed that the magnetoresistance (MR) was increased with the increment of Ni-substituted Zn-Ni ferrites due to spin-dependent scattering. Berchmans et al. [23] investigated the structural and electrical properties of the magnesium-substituted nickel ferrite and found that all samples exhibited semiconductor behavior as the dielectric constantly was decreased with the frequency increment. Notably, during the last several years, the role of nanoparticles as an additive or filler has received significant research interest as sensing applications of MR-based detection methods. Nevertheless, study on this issue is considerably rare. Shabdin et al. [4] found that the application of 33 wt.% of 16 μm graphite particle size in MRE could alter the conductivity and resistivity of MRE subjected to an external applied force. Li et al. [34] fabricated MRE with different concentrations of 5.88 wt.% to 23.81 wt.% graphite and found that samples with higher graphite weight fraction showed higher electrical conductivity and lower decline in resistance. In fact, in previously published data [24,35,36], electrical properties of MRE mainly focused on the resistivity or resistance and contents of fillers. Less consideration has been given to the mechanism of particle interaction with respect to both rheological and electrical properties which are the main concern in practical applications. Although the conductivity of MRE can be improved by increasing the weight fraction of graphite, excessive implementation of particles or fillers in MRE often leads to the decrement of some other properties such as elasticity. Moreover, this makes the processing and dispersion of particles more difficult, which leads to a brittle phase and decrement of the field-dependent modulus of MRE.

The main technical contribution of this work is to fabricate a new MRE associated with the nanosized Ni-Mg cobalt ferrites with different concentrations of Mg, and to experimentally investigate various properties of the proposed MRE including the field-dependent viscoelastic properties and electrical resistance. The nanosized Ni-Mg cobalt ferrites are introduced as a filler in MRE fabrication to enhance magnetic and electric properties of MRE with the least brittleness. It should be noted that to the best of our knowledge, MRE with the nanosized Ni-Mg cobalt ferrites as a filler has never been reported, despite several benefits of the proposed filler. For example, the nanosized Ni-Mg cobalt ferrites exhibit high magnetic remanence and large surface activities [37]. Their properties are believed to demonstrate better magnetic, rheological, as well as electrical properties of MRE. In order to validate the enhanced properties of the proposed MRE, first, three MRE samples are prepared and their morphological observations are recorded. Subsequently, magnetic, field-dependent rheological, and electrical properties are investigated and compared with respect to the different compositions of Mg. It is identified that MRE with the proposed filler can enhance the saturation of magnetization by 3% and the storage modulus and loss factor by 66%. In addition, evaluations show that MRE with a higher content Mg exhibits higher electrical resistance, in the range of 1% to almost 400% in off- and on-state conditions, due to the easier movement of the particles.

2. Methodology

2.1. Samples Preparation

Three types of silicone-based MRE samples were prepared through a mixing process. The CIPs with 70 wt.% of CIP were mixed with 29 wt.% of silicon rubber (SR) (NS 625 A) with the curing agent NS625B (Nippon Steel). A mechanical stirrer was used (multimix high speed dispersed (HSD), WiseStir HT-DX, PMI-Labortechnik GmbH, Lindau, Switzerland) for mixing, for approximately 10 min, until the mixture was visually seen to be homogenous. The conventional co-precipitation method was used to synthesize the nanosized Ni-Mg cobalt ferrites, namely $Co_{0.5}Ni_{0.2}Mg_{0.3}Fe_2O_4$ (A1) and $Co_{0.5}Ni_{0.1}Mg_{0.4}Fe_2O_4$ (A2). The procedures for the synthesized MRE samples, A1 and A2, were based on the previous method reported by Rosnan et al. [37]. The stoichiometric molar amounts of $Ni(NO_3)_2 \cdot 6H_2O$, $Mg(NO_3)_2 \cdot 6H_2O$, $Co(CH_3COO)_2 \cdot 4H_2O$ and $Fe(NO_3)_3 \cdot 9H_2O$ were introduced. The above-mentioned chemicals were purchased from Qrec (New Zealand) and used without further purification. The chemicals were weighted according to the required stoichiometric proportion. Then, as a filler, 1 wt.% of A1 and A2 nanosized Ni-Mg cobalt ferrites, with an average size of 35 to 80 nm, were sonicated using an ultrasonic bath (Clifton, SW6H, UK.) with a frequency of 20 kHz and power of 120 W for 20 min in 100 mL of liquid ethanol (Sigma Aldrich, Malaysia) to prevent agglomeration before mixing with the matrix and CIPs. The MRE and filler were then mixed thoroughly for another 10 min. Then, the curing agent was added, during the mixing with the SR, to the curing agent ratio, with a ratio of 95:5 and continuously stirred for another 1 min. Finally, the MRE samples were cured in the cylinder shape mold for 24 hours. The summary of the fabrication process flow is shown in Figure 1.

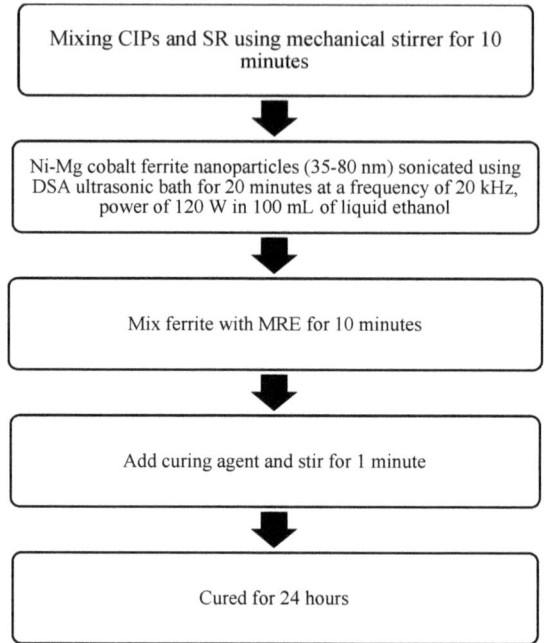

Figure 1. Fabrication flow chart of the proposed magnetorheological elastomer (MRE) samples.

2.2. Samples Characterization

The A1 and A2 morphological image related to the surface and structure of the nanosized Ni-Mg cobalt-ferrite particles were observed using an ultra-high resolution of Field Emission Scanning Electron Microscopy (FESEM), (JSM-7601F, JEOL, Akishima, Tokyo, Japan) at a voltage of 2.0 kV and a magnification of 50,000×. Concurrently, the cross-sections of MRE samples were observed using Scanning Electron Microscopy (SEM), (JSM-IT300LV, JEOL, Akishima, Tokyo, Japan) at a voltage of 5 kV. In order to examine the magnetic properties of all MRE samples, a Vibrating Samples Magnetometer (VSM) (MicroSense, FCM-10, LLC, MA, USA) was used to measure the magnetic properties of the MRE samples. The samples were subjected to the magnetic field in the range of −15 kA/m to +15 kA/m. Meanwhile, the effect of various magnetic fields towards the rheological properties of all MRE samples were evaluated using a rheometer (MCR 302, Anton Paar, Graz, Austria) under oscillation mode at room temperature (25 °C). All samples were subjected to different magnetic fields, ranging from 0 to 5 A (refer to Table 1) and the Tesla unit is used. The experiment is repeated for three times in order to get the consistency of the result obtained. The strain sweep test was carried out by varying the strain from 0.001% to 100%, with a constant frequency of 1 Hz, whereas another test for the current sweep was carried out by varying the magnetic fields from 0 to 5 A, with a constant strain and frequency of 0.01% and 1 Hz, respectively. The electrical resistance properties of all MRE samples (circular shape with a thickness of 1 mm, diameter of 20 mm) were evaluated using a test rig with different forces in order to analyze the electrical resistance behavior. The set-ups were almost the same as the one reported by Shabdin et al. [4]; however, a small modification has been made, in which some load cells have been attached in order to obtain the accuracy of the force, as shown in Figure 2. The test rig consists of data acquisition (DAQ) obtained from the LabVIEW software (6211, National Instruments, Austin, TX, USA). Different loads in the range of 0 to 600 g, increasing in increments of 50 g, were placed on the MRE samples in the absence and presence of a magnetic field. The test was repeated three times in order to obtain the consistency of the electrical resistance of the MRE samples.

Table 1. Magnetic flux density of the MRE samples.

Samples	Magnetic Flux Densities (T)					
	Current (A)					
	0	1	2	3	4	5
MRE	0	0.211	0.412	0.599	0.748	0.854
MRE + A1	0	0.209	0.411	0.598	0.773	0.884
MRE + A2	0	0.214	0.416	0.608	0.774	0.889

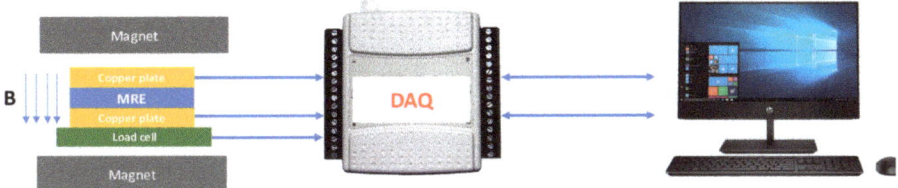

Figure 2. Test rig for the conductivity test.

3. Results and Discussion

3.1. Morphological Properties

The morphologies related to surface and structures of $Co_{0.5}Ni_{0.2}Mg_{0.3}Fe_2O_4$ (A1) and $Co_{0.5}Ni_{0.1}Mg_{0.4}Fe_2O_4$ (A2) at a voltage of 2.0 kV with a magnification of 50,000× were observed in Figure 3.

(a)

Figure 3. Cont.

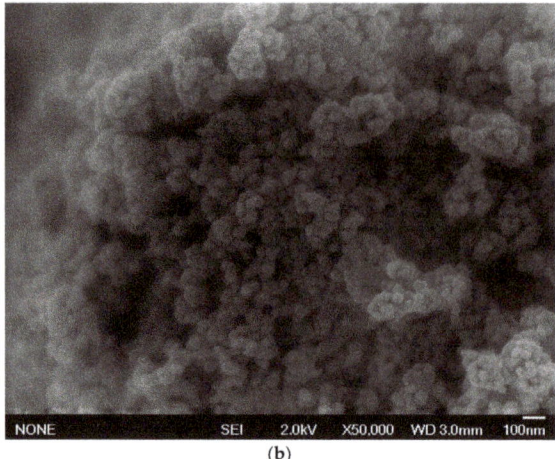

(b)

Figure 3. Field emission scanning electron microscopy (FESEM) images for (**a**) A1 and (**b**) A2 at 50,000× magnification at a voltage of 2.0 kV.

As shown in Figure 3a,b, the A1 and A2 samples exhibit spherical shapes and the nanoparticles are clearly aggregated, which is likely due to strong Van der Waals or magnetostatic interactions between bigger nanoparticles (promoted by the sintering process), a phenomenon which has been reported in previous research by Rosnan et al. [37]. In addition, by using the J-image software, the mean particle sizes observed were in the range of 35 to 80 nm.

Meanwhile, as for the MRE with nanosized Ni-Mg cobalt ferrite, the results of the micrographic analysis for one of the samples (MRE + A2) at different magnifications (1000× and 2500× magnification) are shown below.

Figure 4a displayed the dispersion of particles at 1000× magnification, while Figure 4b showed the dispersion behaviour of particles in an MRE at the higher magnification of 2500×. As demonstrated in Figure 4a,b, the CIPs and nanosized Ni-Mg cobalt-ferrites are randomly dispersed in the silicone rubber matrix. The SEM image also shows that no large aggregation or agglomeration of particles occurred while fabricating the MRE samples, although a small void is detected upon observation of the micrographic analysis.

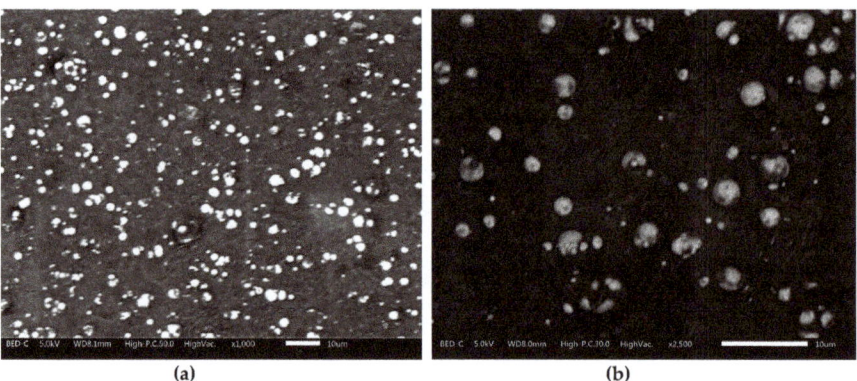

Figure 4. Scanning electron microscopy (SEM) images for MRE + A2 at (**a**) 1000× and (**b**) 2500× magnification.

3.2. Magnetic Properties

The magnetic properties of all MRE samples were measured continuously under magnetic fields in the range of −15 kA/m to 15 k/Am. Figure 5 displays the hysteresis loop of the MRE samples. As shown in Figure 5, the MRE+A1 sample exhibits the highest magnetic saturation, Ms, at 143.92 emu/g, followed by the MRE + A2 sample, at 143.77 Am^2/kg, and control MRE at 140.25 Am^2/kg.

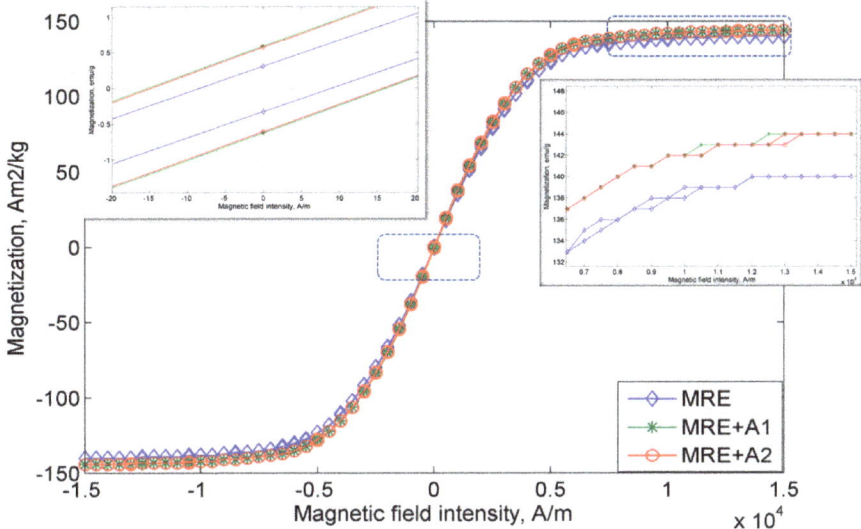

Figure 5. Magnetization curves for all MRE samples.

The summary and comparison of all MRE samples and previous magnetic properties of A1 and A2 are shown in Table 2. The MREs with A1 and A2 exhibit an enhancement of magnetic saturation up to 2% as compared to the control MRE. Due to a higher surface area of the nanosized Ni-Mg cobalt-ferrites, it is believed that this nanoparticle fills the void between the CIPs and resulted in enhancement of the CIPs bonding. In addition, the magnetic remanence, Mr, of MRE with A1 and A2, showed a sharply decreasing value, in the range of 4200% and 4800%, as compared to the nanosized Ni-Mg cobalt-ferrite itself. As a matter of fact, lower remanence is required in any device, particularly in sensor applications as the material in the sensor is highly sensitive to the environment and must have a fast response. As such, the control MRE demonstrates the lowest remanence of 0.32 Am^2/kg as compared to the MRE with A1 and A2. Furthermore, the coercivity, Hc, of the control MRE is also low at 8.70 Oe, while the MRE with A1 and A2 exhibit increases of 81% and 76%, respectively, in comparison to the control MRE. The increase in the coercivity and remanence of MRE with A1 and A2 are believed to be due to the high surface area of the nanosized Ni-Mg cobalt-ferrites, which resulted in easy movement of the nanosized Ni-Mg cobalt-ferrites that were attracted to the magnetic field. By implementing only these types of filler in MRE fabrication, notably, the magnetic properties exhibit larger changes, particularly in terms of remanence and coercivity behavior, which make this type of filler a potential candidate for use in the development of sensors.

Table 2. Summary of the nanosized Ni-Mg cobalt ferrites and all MRE samples.

Samples	MRE	A1 [37]	MRE + A1	A2 [37]	MRE + A2
Ms (Am^2/kg)	140.25	47.34	143.92	54.62	143.77
Mr (Am^2/kg)	0.32	26.41	0.61	29.13	0.59
Hc (Oe)	8.70	621.27	15.71	608.17	15.33

3.3. Sweep Strain

Figure 6 displays the behavior of all MRE samples at different magnetic fields with the increase in the strain sweep. The MRE with A2 exhibited the highest storage modulus at all the magnetic fields applied in the off- and on-state conditions. Meanwhile, the control MRE exhibits a low storage modulus, where the decrease was up to 50% in the off-state condition and in the range of 47–50% in the on-state condition. On the other hand, the MRE with A1 demonstrated the lowest storage modulus, with 0.42 MPa in the off-state condition and a maximum storage modulus of 0.60 MPa in the on-state condition. However, the MREs with A1 and A2 follow the filler rule, in which the filler could contribute to a higher modulus of elasticity and strength than the matrix itself. Thus, in this study, the properties of the storage modulus were found to be changed by changing the weight percent of filler. A higher concentration of Mg in the nanosized Ni-Mg cobalt-ferrites increased the rheological performance of the storage modulus. However, this trend is contradicted for the MRE with A1 as the storage modulus was low as compared to the control MRE. The decrease in the storage modulus in the MRE with A1 might be due to the Ni concentration used in this MRE. As a matter of fact, the magnetic saturation of different ferromagnetic materials, namely Fe, Co and Ni has decreased, which has been reported by previous researchers [38]. It is believed that although the Ni-Mg cobalt-ferrite is able to act as reinforcing filler owing to its large surface area and high aspect ratio, the size of the nanoparticles in the MRE might have an effect on the crosslink density, thus decreasing the storage modulus. Moreover, the decreasing trend of the storage modulus of MRE + A1 might also be due to the broken filler network, which lowers the storage modulus as compared to the control sample. This finding is similar to that observed by previous researchers [39,40]. Meanwhile, for the MRE + A2, it is believed that with the addition of these nanosized Ni-Mg cobalt-ferrites in the MRE, especially in the presence of magnetic field, the hetero-aggregation process occurred between nanosized Ni-Mg cobalt-ferrites and CIPs, thus leading to an enhancement of inter-particle interactions in the MRE. In addition, the void between the CIPs may be filled by the higher surface area and smaller size offered by the nanosized Ni-Mg cobalt-ferrites, therefore, enhancing the interaction between CIPs and nanosized Ni-Mg cobalt-ferrites. A possible mechanism of the above-mentioned phenomenon is shown in Figure 7.

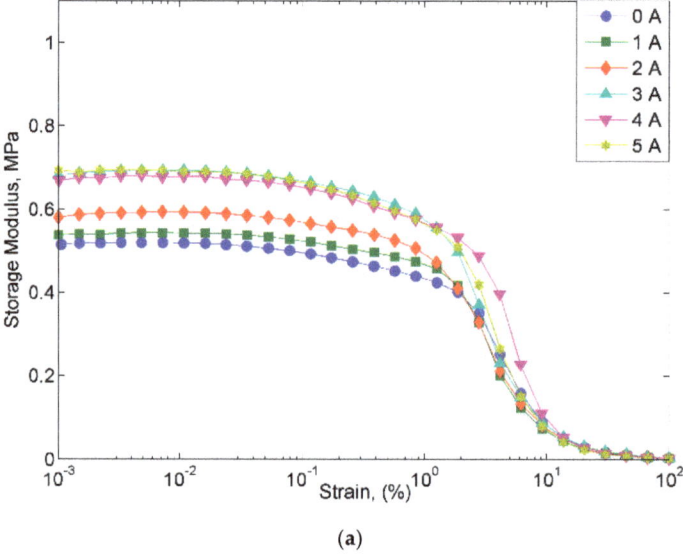

(a)

Figure 6. *Cont.*

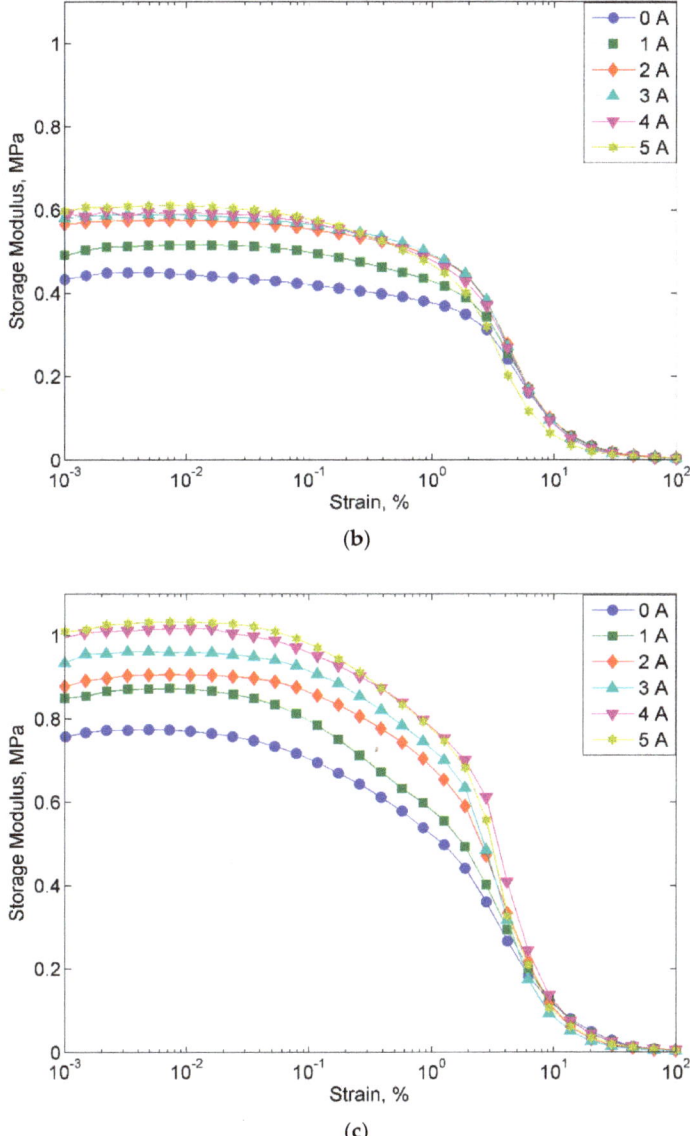

Figure 6. Storage modulus versus strain of all MRE samples (**a**) MRE, (**b**) MRE + A1, and (**c**) MRE + A2.

The mechanism of particle movement is illustrated in Figure 7. For the MRE + A1 and MRE + A2 samples, in the absence of a magnetic field, the CIPs and the nanosized Ni-Mg cobalt-ferrites tend to have a homogenous dispersion. However, when the magnetic field was applied, there are two possible situations that may have occurred. In the first—due to high magnetic saturation of the MREs—the CIPs tend to align according to the magnetic field direction, which is followed by the movement of the nanosized Ni-Mg cobalt-ferrites that will fill the void of the CIPs. In the second possible mechanism—due to the low magnetic moment of the nanosized Ni-Mg cobalt-ferrites and low Mg concentration—they will vibrate and experience slow microscopic movement which shown in Figure 7a. In this situation, the magnetic response as a result of the magnetic field will decrease and produces an obstacle in the matrix, thus reducing the storage modulus capability. Meanwhile, in the MRE + A2 sample, in the presence of a magnetic field, the nanosized Ni-Mg cobalt-ferrites will fill the void between CIPs and strengthen the interaction within the MRE. Moreover, due to high concentration of Mg, this A2 experienced faster microscopic movement which resulted in small void occurred. In this situation, due to the high Mg concentration in the nanosized Ni-Mg cobalt-ferrites, the movement of the nanosized Ni-Mg cobalt-ferrites is increased and the reaction towards the magnetic field is also increased as depicted in Figure 7b. The higher Mg content is assumed to form better bonding due to the increment of compact structure in MRE thus resulted in higher storage modulus. This kind of similar finding has been reported previously by Agarwal et al. [41] in which higher concentration resulted in better mechanical performance.

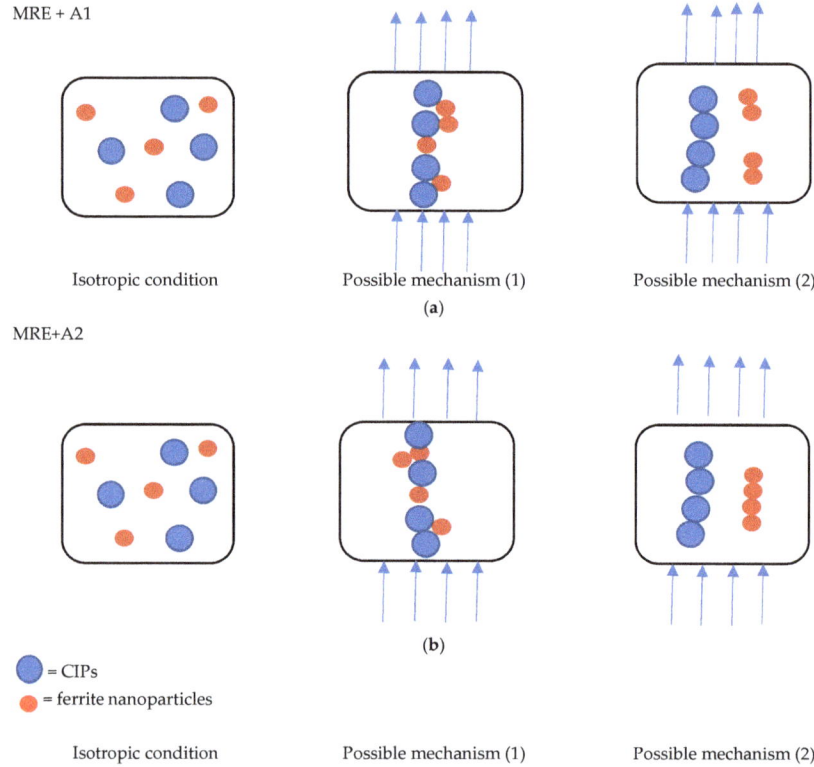

Figure 7. Mechanism of particle movement in the MRE samples (**a**) MRE + A1 and (**b**) MRE + A2.

On the other hand, the loss factor of MRE samples is shown in Figure 8. The samples were subjected to the different magnetic fields, ranging from 0 to 5 A. As shown in Figure 8a–c, it has been identified that the loss factor of all MRE samples is high at a strain of over 1%. In the off-state condition, MRE samples exhibit the lowest loss factor with the increase in the strain as compared to the on-state condition. On the contrary, in the on-state condition, MRE samples depicted a higher loss factor parallel to the increase of the magnetic field. However, the maximum loss factor decreased with the existence of nanosized Ni-Mg cobalt-ferrites in the MRE. This can be attributed to the existence of a magnetic field in which the CIPs and nanosized Ni-Mg cobalt-ferrites tended to vibrate and form a chain-like structure inside the matrix. The higher degree of macroscopic movement of the CIPs resulted in greater particle interaction and internal friction during the vibration. Thus, the energy generated from the vibrating surface of the particles was converted into heat and led to a decrease in the loss factor. Moreover, with the implementation of the nanosized Ni-Mg cobalt-ferrites in the MRE, it is believed that at a higher strain amplitude, the CIPs and nanoparticles are no longer stable due to the breakdown of the filler network. In other words, nanosized Ni-Mg cobalt-ferrites contributed to the increase in the loss factor with the increase in the strain. Therefore, it is important to determine the strain range when applying the MRE samples to the sensors, such as the strain sensor. Moreover, the response time of the sensor needs to be carefully considered if the signal from the MRE is used as a feedback signal in the control system.

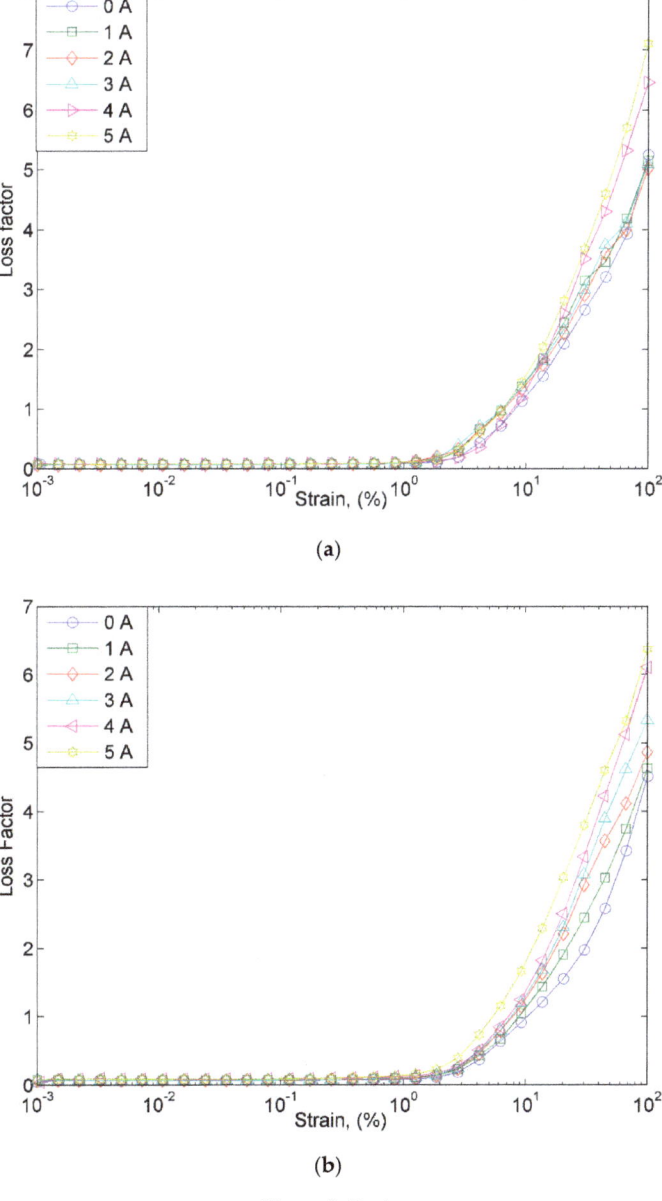

Figure 8. *Cont.*

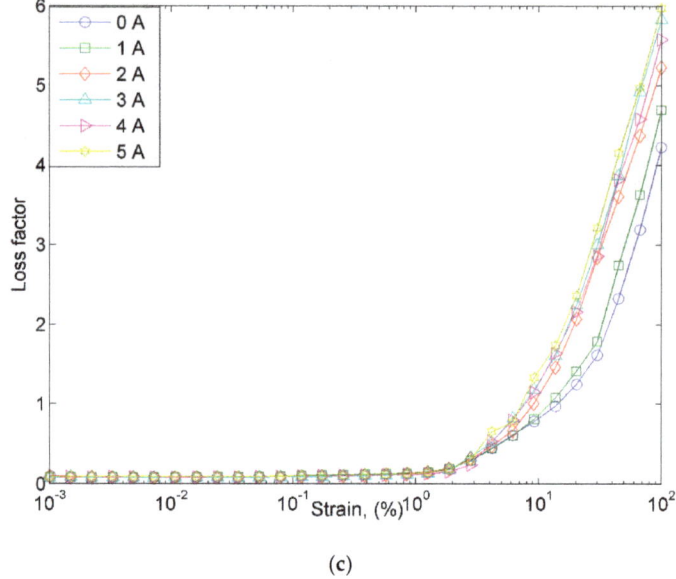

(c)

Figure 8. Loss factor versus strain at various magnetic fields for (**a**) MRE, (**b**) MRE + A1, and (**c**) MRE + A2.

3.4. Sweep Current

Figure 9 presents the storage modulus of the MRE samples at various magnetic flux densities. All MRE samples showed an increasing trend in the storage modulus parallel with the increase of the magnetic field. The summary of the rheological properties of the samples is tabulated in Table 3. The initial storage modulus of MRE + A2 showed a higher value of 0.26 MPa than the control MRE, which was at 0.19 MPa, while, the MRE + A1 exhibited the lowest initial storage modulus of 0.07 MPa. Furthermore, the MRE + A2 depicted the highest storage modulus of 0.82 MPa, followed by the control MRE at 0.50 MPa and MRE + A1 at 0.47 MPa. Both of the MRE samples with A1 and A2 show a higher MR effect—264% and 40% higher, respectively—than the control MRE. It is well known that the enhancement of the rheological properties of MRE depended on the concentration of the magnetic particles, interaction between matrix–filler and also, that of filler–filler. Therefore, the finding in this study revealed that both MREs with nanosized Ni-Mg cobalt-ferrites indicated good compatibility, interaction and bonding between matrix–filler. Due to good adhesion and binding forces between matrix-CIPs and nanosized Ni-Mg cobalt-ferrites, the macromolecular chains are restricted, thus enhancing the storage modulus, which resulted in a higher MR effect. Moreover, the general behavior of the nanosized particles—in which they tend to orient and respond faster than the larger particles—subjected to the magnetic field is also contributed to the increase in the storage modulus and magnetic effect.

Table 3. The initial modulus, absolute (Δ), and relative magnetoresistance (MR) effect.

Samples	G_0 (MPa)	ΔG (MPa)	MR Effect (%)
MRE	0.19	0.30	157.89
MRE + A1	0.07	0.40	571.43
MRE + A2	0.26	0.56	215.38

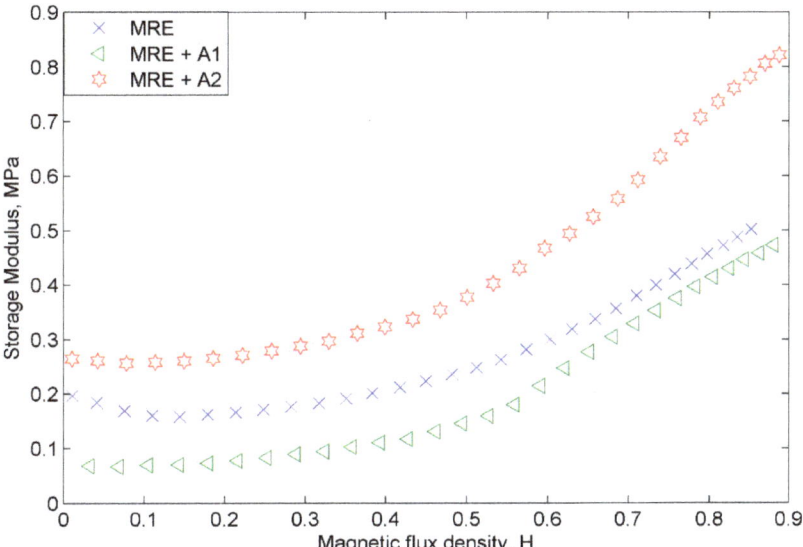

Figure 9. The storage modulus of all MRE samples as a function of magnetic flux density.

3.5. Electrical Resistance of the MRE Samples

The electrical resistance of the MRE samples provides valuable information about the behavior of electric charge carriers, which leads to a good understanding and explanation of the conduction mechanism in MRE with nanosized Ni-Mg cobalt-ferrites. Figure 10 displays the behaviors of MRE with A1 and A2, correlated to different applied weights in the off- and on-state conditions.

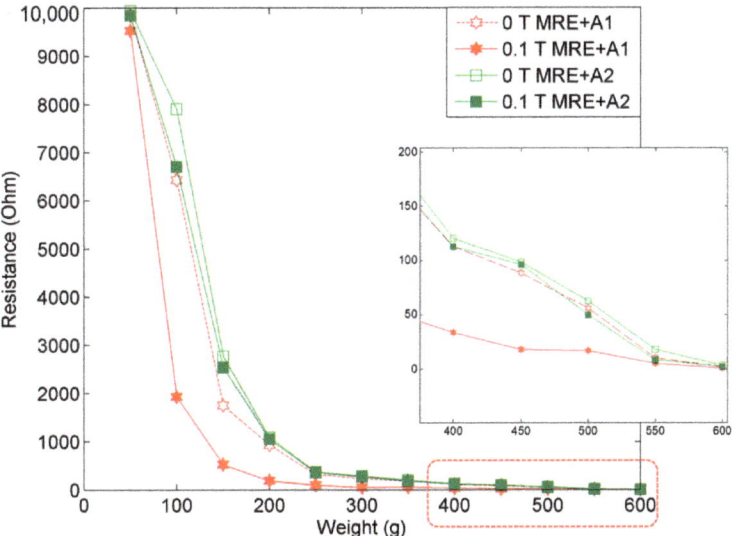

Figure 10. Electrical resistance comparison of MRE samples at the off- (0 T) and the on-state (0.1 T) conditions.

The results obtained by changing the applied weight in the range of 50 to 600 g showed that the electrical resistance decreased at a higher applied weight, which proved that the MRE with nanosized Ni-Mg cobalt-ferrites is capable of exhibiting a significant change towards the electrical resistance of the MRE. Initially, in the off- and on-state conditions—(0 T) and (0.1 T)—the electrical resistances of MRE + A1 and MRE + A2 in the applied weight range from 50 to 600 g at increments of 50 g, showed an exponential decay. The summary of the electrical resistance of both MRE samples is tabulated in Table 4.

Table 4. The electrical resistance of MRE + A1 and MRE + A2 with an increment of weight at the off-state (0 T) and the on-state (0.1 T) conditions.

Samples/Weight (g)	MRE + A1		MRE + A2	
	0 T	0.1 T	0 T	0.1 T
50	9848.63 Ω	9533.67 Ω	9938.91 Ω	9845.05 Ω
100	6431.33 Ω	1929.30 Ω	7906.90 Ω	6698.81 Ω
150	1744.36 Ω	523.21 Ω	2771.35 Ω	2537.91 Ω
200	929.08 Ω	185.84 Ω	1085.38 Ω	1048.48 Ω
250	314.55 Ω	94.25 Ω	369.11 Ω	359.45 Ω
300	232.83 Ω	46.43 Ω	285.56 Ω	268.98 Ω
350	180.45 Ω	54.03 Ω	200.62 Ω	183.40 Ω
400	112.82 Ω	33.67 Ω	120.34 Ω	112.86 Ω
450	88.15 Ω	17.69 Ω	98.26 Ω	96.01 Ω
500	55.68 Ω	16.58 Ω	62.66 Ω	49.30 Ω
550	10.12 Ω	5.06 Ω	17.81 Ω	8.17 Ω
600	1.73 Ω	0.20 Ω	3.02 Ω	2.37 Ω

The results demonstrated that for both the off- and on-state conditions, MRE + A1 exhibited a low electrical resistance as compared to MRE + A2, which implied that MRE + A1 is more sensitive towards the changes in force. In general, in the presence of a magnetic field, the forces between the CIPs and nanosized Ni-Mg cobalt-ferrites increased as the particles tended to form a chain-like structure. In the meantime, due to the existence of the nanosized Ni-Mg cobalt-ferrites in the MRE, the strength of the materials increased. Therefore, the mobility of these particles was restricted and led to a reduction in the electrical resistance.

The mechanism of the particle interactions in the MRE samples in the absence and presence of magnetic and electric fields are shown in Figure 10. As shown in Figure 11a, during the absence of external magnetic and electric fields, the CIPs and the nanosized Ni-Mg cobalt-ferrites tend to disperse randomly in the MRE samples. However, in the presence of a magnetic field, the CIPs tend to form chain-like structures, which can be seen in Figure 11b. In addition, the nanosized Ni-Mg cobalt-ferrites fill the void between the CIPs, which enhanced the attraction force of these particles. Notably, in the presence of magnetic and electric fields (Figure 11c), the CIPs and nanosized Ni-Mg cobalt-ferrites tend to vibrate and form a chain-like structure according to the strength of the magnetic and electrical field applied. Due to the small size of the nanosized Ni-Mg cobalt-ferrites that formed in the magnetic field, the reaction is easier as compared to the CIPs. Hitherto, the CIPs and nanosized Ni-Mg cobalt-ferrites exhibit an increased magnetic moment and lead to an enhancement in the interparticle attraction force. The enhancement of this interparticle attraction force resulted in a decrease in the electrical resistance of the MRE samples, which was proven experimentally and is shown in Figure 10.

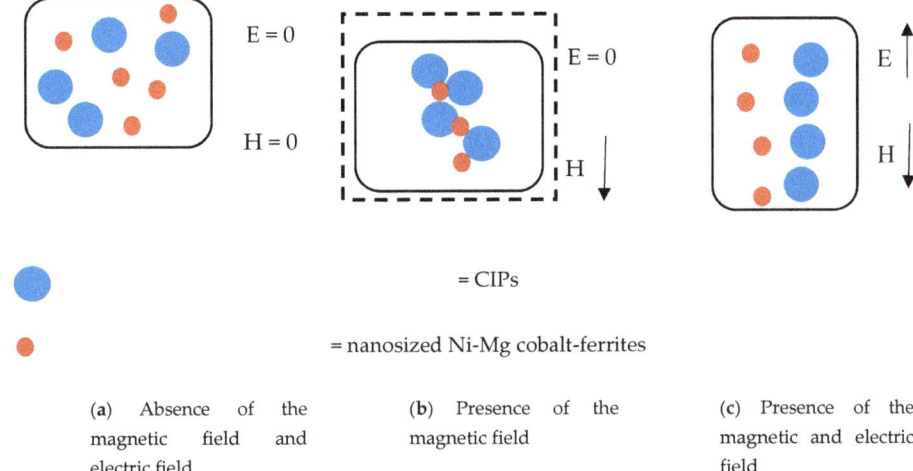

(a) Absence of the magnetic field and electric field

(b) Presence of the magnetic field

(c) Presence of the magnetic and electric field

Figure 11. Mechanism of particles interaction in MRE with the absence and presence of magnetic and electric fields.

4. Conclusions

In this study, MRE samples with and without nanoparticles were successfully fabricated, and the overall properties, related to the magnetic, rheological and electrical resistance properties, were experimentally investigated. It has been shown from the micrographic analysis that the CIPs and nanosized Ni-Mg cobalt-ferrites are randomly dispersed in the MRE samples. More specific results achieved from this work are summarized as follows.

(1) The MRE with nanoparticles as a filler possesses an enhancement of 3% in the saturation of magnetization and decreased the remanence of the nanosized Ni-Mg cobalt-ferrite itself by 4000%.

(2) The storage modulus and loss factor of the MRE with the nanosized Ni-Mg cobalt-ferrites showed an enhancement of almost 66% as compared to the control MRE. The results revealed that the addition of 1.0 wt.% of nanosized Ni-Mg cobalt-ferrites as filler altered and enhanced the interaction of particles in the MRE, and thus led to an increase in the magnetic and rheological properties.

(3) The electrical resistance of the nanosized Ni-Mg cobalt-ferrites decreased with the increase in the applied weight. The MRE with a higher content of Mg in nanosized Ni-Mg cobalt-ferrites exhibited higher electrical resistance in the range of 1% to almost 400% in the off- and on-state conditions due to the easier movement of the particles as a result from the deformation that occurred in the MRE matrix when exposed to an external magnetic field.

It is interesting to note that this work demonstrates that the use of a small concentration (1.0 wt.%) of nanoparticles—which acted as a filler—is capable of altering the magnetic, rheological and electrical resistance behavior of the MRE. This capability directly indicates that MREs with appropriate nanoparticles can be used as actuators or sensors. For example, the stiffness of the flexible structure can be tuned by the MRE (actuator), and the strain (or force) of the flexible structures due to the external force can be measured using the electrical signal generated from the MRE (sensor). Future work will include the investigation of the optimum performance of the electrical resistance of larger Mg concentration and various weights of the nanosized Ni-Mg cobalt-ferrites for MRE samples. In addition, research in terms of the practical aspect of the proposed MREs with the addition of nanoparticles for actuator and/or sensor applications will be undertaken.

Author Contributions: Conceptualization, S.A.A.A., S.A.M., U.U., M.K.S., N.A.Y. and S.-B.C.; Data curation, S.-B.C. and R.M.R.; Formal analysis, S.A.A.A., U.U., R.M.R. and M.K.S.; Funding acquisition, S.A.M., U.U., N.A.Y. and N.A.N.; Investigation, S.A.A.A.; Methodology, S.A.A.A., S.A.M., M.K.S. and R.M.R.; Project administration,

S.A.A.A. and S.A.M.; Resources, S.A.A.A. and N.A.N.; Software, S.A.A.A. and U.U.; Supervision, S.A.M.; Validation, S.A.A.A., S.A.M., U.U., M.K.S., S.-B.C. and R.M.R.; Visualization, S.-B.C.; Writing—original draft, S.A.A.A.; Writing—review and editing, S.A.M., U.U., M.K.S., N.A.Y., N.A.N. and S.-B.C.

Funding: This study was financially supported by the Universiti Teknologi Malaysia under Transdisciplinary Research Grant (Vot No: 06G77), and Professional Development Research University grant (Vot No: 04E02). This work and the APC was also supported and funded by Universiti Teknologi Petronas under Yayasan Universiti Teknologi Petronas-Fundamental Research Grant (YUTP-FRG), Cost Center. 015LC0-135. The research is also partially funded by USAID through Sustainable Higher Education Research Alliances (SHERA) Program- Center for Collaborative (CCR) National Center for Sustainable Transportation Technology (NCSTT) with Contract No. IIE00000078ITB-1, as well as Universitas Sebelas Maret (UNS) through Hibah Mandatory 2019.

Acknowledgments: This study was financially supported by the Universiti Teknologi Malaysia under a Transdisciplinary Research Grant (Vot No: 06G77), and Professional Development Research University grant (Vot No: 04E02). This work was also supported by Universiti Teknologi Petronas under Yayasan Universiti Teknologi Petronas-Fundamental Research Grant (YUTP-FRG), Cost Center. 015LC0-135. The authors also thank the Ministry of Higher Education, Research, and Technology, Indonesia for a mandatory research grant through Hibah WCR (World Class Research) 2019–2020.

Conflicts of Interest: The authors declare no conflict of interest.

References

1. Sun, S.; Deng, H.; Yang, J.; Li, W.; Du, H.; Alici, G. Performance evaluation and comparison of magnetorheological elastomer absorbers working in shear and squeeze modes. *J. Intell. Mater. Syst. Struct.* **2015**, *26*, 1757–1763. [CrossRef]
2. Wahab, N.A.A.; Mazlan, S.A.; Ubaidillah, U.; Kamaruddin, S.; Ismail, N.I.N.; Choi, S.B.; Sharif, A.H.R. Fabrication and investigation on field-dependent properties of natural rubber based magneto-rheological elastomer isolator. *Smart Mater. Struct.* **2016**, *25*, 107002. [CrossRef]
3. Yang, J.; Sun, S.S.; Du, H.; Li, W.H.; Alici, G.; Deng, H.X. A novel magnetorheological elastomer isolator with negative changing stiffness for vibration reduction. *Smart Mater. Struct.* **2014**, *23*, 105023. [CrossRef]
4. Shabdin, M.K.; Rahman, M.A.A.; Mazlan, S.A.; Ubaidillah; Hapipi, N.M.; Adiputra, D.; Aziz, S.A.A.; Bahiuddin, I.; Choi, S.B. Material characterizations of gr-based magnetorheological elastomer for possible sensor applications: Rheological and resistivity properties. *Materials* **2019**, *12*, 391. [CrossRef] [PubMed]
5. Wang, Y.; Zhang, X.; Oh, J.; Chung, K. Fabrication and properties of magnetorheological elastomers based on CR/ENR self-crosslinking blends. *Smart Mater. Struct.* **2015**, *24*, 095006. [CrossRef]
6. Tian, T.F.; Zhang, X.Z.; Li, W.H.; Alici, G.; Ding, J. Study of PDMS based magnetorheological elastomers. *J. Phys. Conf. Ser.* **2013**, *412*, 012038. [CrossRef]
7. Jung, H.S.; Kwon, S.H.; Choi, H.J.; Jung, J.H.; Kim, Y.G. Magnetic carbonyl iron/natural rubber composite elastomer and its magnetorheology. *Compos. Struct.* **2016**, *136*, 106–112. [CrossRef]
8. Zaborski, M.; Pietrasik, J.; Masłowski, M. Elastomers containing fillers with magnetic properties. *Solid State Phenom.* **2009**, *154*, 121–126. [CrossRef]
9. Koo, J.H.; Dawson, A.; Jung, H.J. Characterization of actuation properties of magnetorheological elastomers with embedded hard magnetic particles. *J. Intell. Mater. Syst. Struct.* **2012**, *23*, 1049–1054. [CrossRef]
10. Kramarenko, E.Y.; Chertovich, A.V.; Stepanov, G.V.; Semisalova, A.S.; Makarova, L.A.; Perov, N.S.; Khokhlov, A.R. Magnetic and viscoelastic response of elastomers with hard magnetic filler. *Smart Mater. Struct.* **2015**, *24*, 035002. [CrossRef]
11. Jin, Q.; Xu, Y.G.; Di, Y.; Fan, H. Influence of the particle size on the rheology of magnetorheological elastomer. *Mater. Sci. Forum* **2014**, *809–810*, 757–763. [CrossRef]
12. Hapipi, N.; Aziz, S.A.A.; Mazlan, S.A.; Ubaidillah; Choi, S.B.; Mohamad, N.; Khairi, M.H.A.; Fatah, A.Y.A. The field-dependent rheological properties of plate-like carbonyl iron particle-based magnetorheological elastomers. *Results Phys.* **2019**, *12*, 2146–2154. [CrossRef]
13. Borin, D.; Kolsch, N.; Stepanov, G.; Odenbach, S. On the oscillating shear rheometry of magnetorheological elastomers. *Rheol. Acta* **2018**, *57*, 217–227. [CrossRef]
14. Tong, Y.; Dong, X.; Qi, M. Improved tunable range of the field-induced storage modulus by using flower-like particles as the active phase of magnetorheological elastomers. *Soft Matter* **2018**, *14*, 3504–3509. [CrossRef]
15. Sapouna, K.; Xiong, Y.P.; Shenoi, R.A. Dynamic mechanical properties of isotropic/anisotropic silicon magnetorheological elastomer composites. *Smart Mater. Struct.* **2017**, *26*, 115010. [CrossRef]

16. Naimzad, A.; Hojjat, Y.; Ghodsi, M. Comparative study on mechanical and magnetic properties of porous and nonporous film-shaped magnetorheological nanocomposites based on silicone rubber. *Int. J. Innov. Sci. Mod. Eng.* **2014**, *2*, 11–20.
17. Schubert, G.; Harrison, P. Large-strain behaviour of magneto-rheological elastomers tested under uniaxial compression and tension, and pure shear deformations. *Polym. Test.* **2015**, *42*, 122–134. [CrossRef]
18. Kaur, R.; Hasan, A.; Iqbal, N.; Alam, S.; Saini, M.K.; Raza, S.K. Synthesis and surface engineering of magnetic nanoparticles for environmental cleanup and pesticide residue analysis: A review. *J. Sep. Sci.* **2014**, *37*, 1805–1825. [CrossRef]
19. Allia, P.; Barrera, G.; Tiberto, P.; Nardi, T.; Leterrier, Y.; Sangermano, M. Fe_3O_4 nanoparticles and nanocomposites with potential application in biomedicine and in communication technologies: Nanoparticle aggregation, interaction, and effective magnetic anisotropy. *J. Appl. Phys.* **2014**, *116*, 113903. [CrossRef]
20. Laurent, S.; Forge, D.; Port, M.; Roch, A.; Robic, C.; Vander Elst, L.; Muller, R.N. magnetic iron oxide nanoparticles: Synthesis, stabilization, vectorization, physicochemical characterizations, and biological applications. *Chem. Rev.* **2008**, *108*, 2064–2110. [CrossRef]
21. Mordina, B.; Tiwari, R.K.; Setua, D.K.; Sharma, A. Superior elastomeric nanocomposites with electrospun nanofibers and nanoparticles of $CoFe_2O_4$ for magnetorheological applications. *RSC Adv.* **2015**, *5*, 19091–19105. [CrossRef]
22. Akther Hossain, A.K.M.; Seki, M.; Kawai, T.; Tabata, H. Colossal magnetoresistance in spinel type Zn1−xNixFe2O4. *J. Appl. Phys.* **2004**, *96*, 1273–1275. [CrossRef]
23. John Berchmans, L.; Kalai Selvan, R.; Selva Kumar, P.; Augustin, C. Structural and electrical properties of Ni1−xMgxFe2O4 synthesized by citrate gel process. *J. Magn. Magn. Mater.* **2004**, *279*, 103–110. [CrossRef]
24. Tian, T.F.; Li, W.H.; Deng, Y.M. Sensing capabilities of graphite based MR elastomers. *Smart Mater. Struct.* **2011**, *20*, 025022. [CrossRef]
25. Ahmad Khairi, M.H.; Mazlan, S.A.; Ubaidillah; Ku Ahmad, K.Z.; Choi, S.B.; Abdul Aziz, S.A.; Yunus, N.A. The field-dependent complex modulus of magnetorheological elastomers consisting of sucrose acetate isobutyrate ester. *J. Intell. Mater. Syst. Struct.* **2017**, *28*, 1993–2004. [CrossRef]
26. Wang, Y.; Xuan, S.; Dong, B.; Xu, F.; Gong, X. Stimuli dependent impedance of conductive magnetorheological elastomers. *Smart Mater. Struct.* **2016**, *25*, 025003. [CrossRef]
27. Aziz, S.A.A.; Mazlan, S.A.; Ismail, N.I.N.; Ubaidillah; Choi, S.B.; Nordin, N.A.; Mohamad, N. A comparative assessment of different dispersing aids in enhancing magnetorheological elastomer properties. *Smart Mater. Struct.* **2018**, *27*, 117002. [CrossRef]
28. Poojary, U.R.; Hegde, S.; Gangadharan, K.V. Experimental investigation on the effect of carbon nanotube additive on the field-induced viscoelastic properties of magnetorheological elastomer. *J. Mater. Sci.* **2018**, *53*, 4229–4241. [CrossRef]
29. Li, R.; Sun, L.Z. Dynamic viscoelastic behavior of multiwalled carbon nanotube–reinforced magnetorheological (MR) nanocomposites. *J. Nanomech. Micromech.* **2014**, *4*, A4013014. [CrossRef]
30. Aziz, S.A.A.; Ubaidillah, U.; Mazlan, S.A.; Ismail, N.I.N.; Choi, S. Implementation of functionalized multiwall carbon nanotubes on magnetorheological elastomer. *J. Mater. Sci.* **2018**, *53*, 10122–10134. [CrossRef]
31. Abdul Aziz, S.A.; Mazlan, S.A.; Nik Ismail, N.I.; Choi, S.B.; Ubaidillah, U.; Yunus, N.A.B. An enhancement of mechanical and rheological properties of magnetorheological elastomer with multiwall carbon nanotubes. *J. Intell. Mater. Syst. Struct.* **2017**, *28*, 3127–3138. [CrossRef]
32. Padalka, O.; Song, H.J.; Wereley, N.M.; Filer II, J.A.; Bell, R.C. Stiffness and damping in Fe, Co, and Ni nanowire-based magnetorheological elastomeric composites. *IEEE Trans. Magn.* **2010**, *46*, 2275–2277. [CrossRef]
33. Yu, M.; Zhu, M.; Fu, J.; Yang, P.A.; Qi, S. A dimorphic magnetorheological elastomer incorporated with Fe nano-flakes modified carbonyl iron particles: Preparation and characterization. *Smart Mater. Struct.* **2015**, *24*, 115021. [CrossRef]
34. Li, W.; Kostidis, K.; Zhang, X.; Zhou, Y. Development of a Force Sensor Working with MR Elastomers. In Proceedings of the 2009 IEEE/ASME International Conference on Advanced Intelligent Mechatronics, Singapore, 14–17 July 2009; pp. 233–238.
35. Tian, T.F.; Li, W.H.; Alici, G. Study of magnetorheology and sensing capabilities of MR elastomers. *J. Phys. Conf. Ser.* **2013**, *412*, 012037. [CrossRef]

36. Li, W.; Tian, T.; Du, H. Sensing and rheological capabilities of MR elastomers. In *Smart Actuation and Sensing Systems - Recent Advances and Future Challenges*; InTech: London, UK, 2012.
37. Rosnan, R.M.; Othaman, Z.; Hussin, R.; Ati, A.A.; Samavati, A.; Dabagh, S.; Zare, S. Effects of Mg substitution on the structural and magnetic properties of $Co_{0.5}Ni_{0.5-x}Mg_xFe_2O_4$ nanoparticle ferrites. *Chinese Phys. B* **2016**, *25*, 047501. [CrossRef]
38. Andersson, S. Spin-Diode Effect and Thermally Controlled Switching in Magnetic Spin-Valves. Ph.D. Thesis, KTH Applied Physics, Stockholm, Sweden, March 2012.
39. Liu, T.; Esposito, C.I.; Burket, J.C.; Wright, T.M. Crosslink density is reduced and oxidation is increased in retrieved highly crosslinked polyethylene TKA tibial inserts. *Clin. Orthop. Relat. Res.* **2017**, *475*, 128–136. [CrossRef]
40. Mujtaba, A.; Keller, M.; Ilisch, S.; Radusch, H.J.; Thurn-Albrecht, T.; Saalwächter, K.; Beiner, M. Mechanical properties and cross-link density of styrene–butadiene model composites containing fillers with bimodal particle size distribution. *Macromolecules* **2012**, *45*, 6504–6515. [CrossRef]
41. Agarwal, G.; Patnaik, A.; kumar Sharma, R.; Agarwal, J. Effect of stacking sequence on physical, mechanical and tribological properties of glass-carbon hybrid composites. *Friction* **2014**, *2*, 354–364. [CrossRef]

 © 2019 by the authors. Licensee MDPI, Basel, Switzerland. This article is an open access article distributed under the terms and conditions of the Creative Commons Attribution (CC BY) license (http://creativecommons.org/licenses/by/4.0/).

Review

Magnetorheological Elastomers: Fabrication, Characteristics, and Applications

Sung Soon Kang [1], Kisuk Choi [2], Jae-Do Nam [2] and Hyoung Jin Choi [1,*]

[1] Department of Polymer Science and Engineering, Inha University, Incheon 22212, Korea; 22191263@inha.edu
[2] Department of Polymer Science and Engineering, School of Chemical Engineering, Sungkyunkwan University, Suwon 16419, Korea; kisuk929@skku.edu (K.C.); jdnam@skku.edu (J.-D.N.)
* Correspondence: hjchoi@inha.ac.kr; Tel.: +82-32-860-7486

Received: 31 August 2020; Accepted: 13 October 2020; Published: 15 October 2020

Abstract: Magnetorheological (MR) elastomers become one of the most powerful smart and advanced materials that can be tuned reversibly, finely, and quickly in terms of their mechanical and viscoelastic properties by an input magnetic field. They are composite materials in which magnetizable particles are dispersed in solid base elastomers. Their distinctive behaviors are relying on the type and size of dispersed magnetic particles, the type of elastomer matrix, and the type of non-magnetic fillers such as plasticizer, carbon black, and crosslink agent. With these controllable characteristics, they can be applied to various applications such as vibration absorber, isolator, magnetoresistor, and electromagnetic wave absorption. This review provides a summary of the fabrication, properties, and applications of MR elastomers made of various elastomeric materials.

Keywords: magnetorheological; elastomer; magnetic particle; viscoelastic; rheological

1. Introduction

As the increasing trend of better and comfortable lifestyle has led to growing demand for both new technologies and materials, advanced smart and intelligent functional materials have been receiving a large attentions in recent years [1]. The smart materials are those controllable with external environments such as electric or magnetic field, mechanical stress, heat, and light [2].

Among these, magnetorheological (MR) materials become one of the most important smart materials in terms of their huge industrial potentials. They are classified as a functional smart material possessing tunable rheological and viscoelastic properties such as yield stress, shear stress, dynamic moduli, and damping property when an external magnetic field is applied [3–7]. Note that while their electrical analogue electrorheological (ER) materials have been also extensively investigated [8], MR materials prevail ER materials regarding both their scientific investigation and applications due to superior performance characteristics of the MR materials [9].

MR materials in general include several different systems, depending on their media employed such as MR fluid (MRF), MR foams, MR gels, and MR elastomer (MRE) [4]. Historically, MRFs, colloidal suspensions that consist of small soft-magnetic particles with extremely small hysteresis, high magnetic permeability and high saturation magnetization suspended in non-magnetic liquids [10] have been widely studied. Dispersed soft-magnetic particles enhance the apparent shear viscosity and yield stress of the MR suspensions [11], depending on both magnetic field strength applied [12] and concentration of micron-sized magnetic particles [13,14]. They have been extensively adopted in various applications such as MR brake, MR valve, MR mount, MR clutch, and MR damper [15]. The versatility of their wide applications is caused by noiseless operation, rapid field responsiveness of MRFs relative insensitivity to small quantities of contaminants or dust, and easy control [16]. On the other hand, despite that MRFs are possessing significant merits with many potential applications, their disadvantages include sealing issues due to the leakage of the medium liquid, contaminating environment, and sedimentation of the particles [17]. Hence these drawbacks limit to the further expansion of its engineering applications.

Contrary to the MRFs, MREs are the hard (particles) and soft (pristine matrix) hybrid composite smart elastomeric materials [18]. Therefore, unlike MRFs, solid-state MREs overcome the disadvantages of MRFs. MRE contains unique mechanical properties in such a way that their viscoelastic properties change under an applied magnetic field. This behavior is very similar to the MRF, but MRE is a more like solid equivalent [19]. In addition, the MRE is generally composed of a material that contains a rubber matrix, minor additives and magnetic carbonyl iron (CI) particles. In such a material, the coalescence of CI particles occurs within an applied magnetic field, which attributes to the hardness, enhanced elastic modulus, and shear modulus [20]. The typical rubber matrix includes silicone rubber (SR) [21–23], polybutadiene rubber [24,25], nitrile rubber [26], and polyurethane (PU) rubber [27,28] and others. Traditionally, elastomers are nonmetallic materials, being used in wide engineering applications such as seals, gaskets, and tires [29]. Correspondingly, the MREs are mainly used in three main categories in past decades, which are sensing devices, actuators and vibration, and noise control [30]. In recent studies, the MRE is also adapted in biomedical engineering field where it is used as a soft material for magnetic-elastic soft robot [31,32]. The tunable viscoelastic properties of MREs have also overpowered the mobility limitation of small-scale robot due to texture or different material in an unstructured environment [31]. Hence, according to these studies, MREs have drawn dramatic interests in various engineering applications.

On the other hand, MREs can be classified into two different groups, which are isotropic and anisotropic MREs, based on mechanisms of magnetically polarized particle configuration in the MREs. The polarized particles are uniformly suspended in an isotropic MRE, so that the MRE shows homogenous physical behaviors in every direction. For an anisotropic MRE, the magnetic particles are aligned along with the input magnetic field direction during manufacturing process, which leads to the perpendicular direction of a flat MRE specimen. In many studies, the shear direction that is perpendicular to the CI particles' alignment was observed when a rheological measurement of anisotropic MRE was conducted [33–35]. Tian and Nakano [36] proposed the fabrication process of anisotropic MRE with 45° CI particles' (less than 60 μm depending on their shape) arrangement in various silicone oil concentrations, which enhanced the storage modulus (G′). In addition, the combination of isotropic/anisotropic MRE has examined to adjust damping capability and zero-field dynamic stiffness of silicone based MRE [37]. According to these studies, MRE has raised great attention in engineering fields. On the other hand, along with the CI particles, various magnetic particles such as $CoFe_2O_4$ (less than 12 nm) [38], Ni (10 μm) [39], $FeCO_3$ (less than 30 nm) [40], and industrial waste nickel zinc ferrite (less than 2 μm) [41] have been also employed. Siti et al. [42] fabricated MREs using CI and nano-sized Ni-Mg cobalt ferrites (35 to 80 nm) based on SR. Even though, both MRF and MRE possess similar magnetic field properties, the main distinction is that their operation period is dependent within two types of materials. The interesting point is that MRE tends to operate in pre-yield regime [43,44], where MRF is typically operated in a flow regime or post-yield shear regime [45,46]. Hence, the magnetic field performance of MRF is mainly obtained from yield stress while MRE is characterized by field-dependent modulus [47].

This article reviews a recent trend on preparation, characteristics, and applications of smart MRE materials. While the microstructures with isotropic or anisotropic MREs are tested by scanning electron microscopy (SEM) and the magnetic property of the magnetic particles used was analyzed by VSM, the mechanical characteristics of MREs such as tensile, Payne, and loss factor have been conducted. In addition, the rheological properties of MREs including dynamic, amplitude sweep, frequency, sweep, and creep test are also illustrated.

2. MRE Materials

2.1. Fabrication

The MRE is generally treated as solid analogues of MRF, being consisted of magnetized particles in polymer medium. The manufacturing procedure of MRE is similar to that of the ordinary rubbers. The materials include soft-magnetic particles, elastomers, and additives. The MREs can be categorized into two classes, which are isotropic and anisotropic upon the curing condition whether it requires applied magnetic field or not. The fabrication process of MRE includes three steps: Mixing, curing, and magnetic particle alignment under an input magnetic field as presented in Figure 1. The isotropic MRE is without a magnetic field while an anisotropic MRE requires an applied magnetic field. For example, Puente-Córdova et al. [48] prepared MREs with two different concentration (20 and 30 wt%) of CI, silicone oil, and tin catalyst. The magnetic particles were dispersed in silicone oil and mixed with SR and catalyst at room temperature. For such high viscosity of the solution requires high mixing process to minimize the sedimentation of the added magnetic particles. The curing process is usually conducted within 12 h under vacuum environment to avoid porosity. The homogenous mixture was exposed to a magnetic flux density of 7 mT only for anisotropic reinforced samples. It is crucial to note that the mixture reaches the semi-solid state, which mitigate magnetic particle sedimentation after 20 min of curing process.

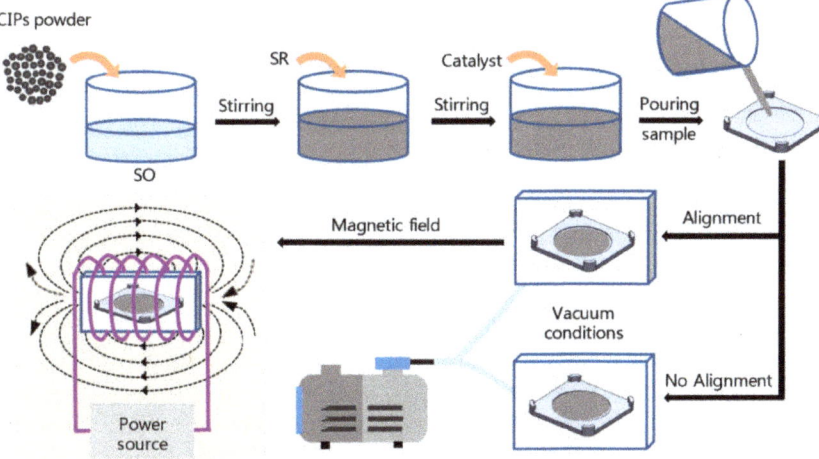

Figure 1. Schematics of manufacturing step of both isotropic and anisotropic magnetorheological elastomer (MREs) (stirring, mixing, and alignment with or without magnetic field) [48].

2.2. Soft-Magnetic Particles

MREs are fabricated by adding high magnetic permeability particles to a viscoelastic material [49]. As such, the magnetic particles added to the MR elastomer play an important role in the MR effect. In general, the conditions of good magnetic particle for MR materials require high saturation magnetization, low remnant magnetization, and high, short-term inter-particle attraction. Of the many magnetically permeable particles, the CI which was discovered by BASF in 1925 produced by thermal decomposition of iron pentacarbonyl (Fe(CO)$_5$) is the most studied and used as MR particle [50–53]. The reason is that as one of the soft-magnetics, it has the advantage of high saturation magnetization, easy magnetization and demagnetization, and no magnetic hysteresis. Because of these properties, it has been used in many fields such as MRFs [54], MR foams [55], and MR greases [56]. CI particles are soft magnetic microparticles being categorized in two groups based on their shape, which are spherical and flake-typed with a size of 1–7 μm and 5–50 μm, respectively [50,54]. In the case of MRF, due to the sedimentation problem caused by high density difference between medium oil and CI particles, their sedimentation problem has been improved by making their density relatively low by being coated with several organic materials. For example, coating polymers such as poly(methyl methacrylate) (PMMA) [6], PS [7], polyaniline [44], and polycarbonate [57] not only reduce the density of CI particles but also prevent chemical oxidation of CI surfaces. However, unlike MRFs, particle sedimentation is not a problem for MREs. The problem with MREs is the compatibility between the magnetic particles and the elastomeric medium. The reason is that CI particles are hydrophilic, whereas elastomer matrices are generally hydrophobic. Therefore, many studies have been conducted to increase the surface compatibility of the CI matrix with the elastomer matrix.

Among various surface modification methods, the treatment of CI particles with silane coupling chemicals is known to be a very economical and efficient way to increase the affinity between the CI particles and the medium [58]. An et al. [58] studied (3-aminopropyl) triethoxysilane (APTES) modified CI particles to improve compatibility with the elastomer matrix (Figure 2b). They fabricated a natural rubber based anisotropic MRE using CI particles surface modified with APTES. And they measured the properties of MRE according to the surface treatment. In order to coat the CI surface with APTES, hydrochloric acid was used for the pre-treatment. This enables the CI surface to be coated by activating the OH group to polymerize APTES onto the CI surface. Figure 2b is a SEM image, showing that the surface of the CI has a rough surface by APTES coating. When CI is coated with APTES in this way, it has NH_2 group on the surface of CI, making CI particles to have a good affinity with natural rubber.

Figure 2. SEM image of (**a**) pure carbonyl iron (CI), (**b**) CI/(3-aminopropyl) triethoxysilane (APTES) [58], (**c**) CI/poly(methyl methacrylate) (PMMA) [59], (**d**) CI/poly(glycidyl methacrylate) (PGMA) [60], (**e**) CI/poly(fluorostyrene) [61], (**f**) CI/poly(trimethylsilyloxyethyl methacrylate) (PHEMATMS) [62].

Many studies have been carried out on coating CI particles with PMMA and applying them to MRFs [6,63]. The PMMA-coated core–shell structured CI particles have a lower density than neat CI particles, resulting in better dispersion stability in the fluid. Li et al. [59] have applied CI/PMMA core–shell particles to MREs rather than MRFs (Figure 2c). The rough surface represents how the CI particles were covered by PMMA particle. Fabricating MREs with CI/PMMA core–shell particle results in lower Payne effects by increasing bond strength between matrix and particles. CI particles were coated with PMMA by emulsion polymerization. First, the surface of the CI particles was activated with acetic acid and then dispersed in distilled water. Sodium lauryl sulfonate is used as a stabilizer to suspend the CI particles, and the CI surface is coated with PMMA using MMA as a monomer and ammonium persulfate as an initiator. As shown in the Figure 2c, PMMA coated CI surface can be confirmed. Using the MRE made of CI/PMMA core–shell particles, the compatibility between particle and matrix is increased and the relative motion between particle and medium is reduced, resulting in smaller and more stable loss factors.

Kwon et al. [60] coated poly(glycidyl methacrylate) (PGMA) on CI particles and applied them to MREs and studied their properties (Figure 2d). The surface of the CI/PGMA particles was observed to be rougher than those of the pristine CI particles. The process of producing core–shell particles using the dispersion polymerization method has the advantage of being simple in a single step process. Due to the hydrophobicity of PGMA, coating PGMA on magnetic particles has the advantage of increasing chemical affinity between the CI particles and the elastomer. PGMA coated CI particles increase the hardness of the MRE matrix, resulting in lower MR effects than MREs made from pure CI. PGMA coatings enhance the bond strength between the CI particles and the elastomeric medium, reducing the loss tangent.

Fuchs et al. [61] researched the coating of poly(fluorostyrene) on CI particles with the ATRP method and applied them to MREs (Figure 2e). By coating poly(fluorostyrene) on the CI surface, it was possible to prevent the oxidation of magnetic particles, one of the potential problems with MRE. It prevents mechanical property from decreasing due to the rapid oxidation of magnetic particles, making it possible to apply for a longer time as a vibration isolator.

Martin et al. [62] reported the characteristics of MREs by coating poly(trimethylsilyloxyethyl methacrylate) (PHEMATMS) on the surface of CI particles, in which the CI-g-PHEMATMS particles showed about 6.2% reduction in magnetization than neat CI particles due to the 10–20 nm coating thickness. However, it was possible to improve thermo-oxidation stability and anti-acid/corrosion properties through polymer coating. In the case of neat CI, wettability is low, which brings a strong reinforcing effect to the MRE. However, when CI is coated with PHEMATMS, mobility is improved inside the elastomer matrix, creating a plasticizing effect. As a result, MRE with CI-g-PHEMATMS particles shows lower G' and higher G" than MRE with neat CI. This results in a higher damping factor. In addition, the increase in particle mobility increases the relative MR effect, which is expected to be a more practical application.

In addition to CI particles, the effects of many types of magnetic particles such as $CoFe_2O_4$ [38], Ni [39], $FeCO_3$ [40], and manganese zinc ferrite [41] have been studied. Siti et al. [42] fabricated MREs using CI and nano-sized Ni-Mg cobalt ferrites based on SR, and studied various properties including field-dependent viscoelastic characteristics of MREs. The experimental results showed that the G' and loss factor were increased in MRE containing Ni-Mg cobalt ferrite nanoparticles. This means that Ni-Mg cobalt ferrite nanoparticles increase the interaction between MRE and particles. On the other hand, Nordalila et al. [64] manufactured MRE using industrial waste nickel-zinc ferrite. Experiments were conducted to compare the degree of swelling according to the content of nickel-zinc ferrite. The thermal analysis of MRE and the MR effect of anisotropic and isotropic MREs were also performed.

Recently, in addition to these soft magnetic particles, studies have been made on MREs using hard magnetic fillers [21,65–67]. Conventional soft magnetic particle-based MREs aimed at maximizing the MR effect. But the MR elastomer fabricated with hard magnet behaves like a flexible permanent magnet. Due to these characteristics, hard magnet-based elastomers have been applied to other applications. When an input magnetic field is present, hard magnetic particles cause rotational motion, which can cause motion and force like a magnetic field-controlled actuator. Koo et al. [21] studied the actuation properties of hard magnet MRE (H-MRE) using barium hexaferrite, strontium ferrite, samarium cobalt, and neodymium magnet. It offers the possibility that MREs with a hard magnet base can be used as magnetically controlled actuators. Therefore, it can be noted that MREs were studied by applying various magnetic particles from soft magnetic to a hard magnet.

2.3. Types of Elastomer

2.3.1. Silicone Rubber

Unlike other organic rubbers, the silicone rubber (SR) has both organic and inorganic properties due to Si-O bonding. Because of these characteristics, heat resistance, its electrical insulation, abrasion resistance, and chemical stability are superior to general organic rubber [24,26]. Therefore, it has been used in many industrial areas including food processing and medical devices. Because of these advantages, it has also been used as a matrix of MREs. Among many SRs, polydimethylsiloxanes (PDMSs) were adopted as the matrix of MRE. Due to the chemical neutrality and stability of PDMS system, it has also a good adhesive property with metal. PDMS rubbers have the advantage of being able to cure at low temperatures and fast curing at high temperatures. In addition, the precursor has a low shear viscosity, making it easy to manufacture MREs [27,68,69]. Xu et al. [70] manufactured a PDMS based MRE and studied the movement mechanism of magnetic particles in the MRE according to the magnetic field. Li and Nakano [71] have reported the mechanical and MR effects of the CI and PDMS matrix base MRE.

2.3.2. Natural Rubber

Natural rubber (NR) is a polymer of isoprene made by solidifying latex obtained from the Hevea brasiliensis tree, being generally composed of cis-constituted C_5H_8 units (isoprene) with one double bond in each repeat unit. NRs as unsaturated elastomers have superior characteristics such as high strength, better resistance, and elongation at break [72]. However, due to the presence of double bonds in the chain, they are very sensitive to heat and oxidation. The improvement in rubber elasticity and strength is usually obtained by a vulcanization process in the presence of sulfur, accelerators and other compounding components, creating a three-dimensional network. NR, which has undergone through this vulcanization process, has higher mechanical performance than other rubbers, making it more suitable for practical applications [30]. Chen et al. [73] used NR to study the MRE properties according to temperature, plasticizer, and iron particle content. They observed that NR had better mechanical properties than SR, and as CI contents increased, the shear modulus of MRE increased, resulting in an increase in MRE performance. However, as CI contents increased, tensile strength and angle tear strength of MRE decreased. Aziz et al. [74] studied the effects of dispersing agents such as naphthenic oil, light mineral oil, and epoxidized palm oil on MR rubber based on NR. Compared to conventional petroleum-based dispersing aids, these eco-friendly dispersing aids, also increased mechanical properties such as MRE's magnetic behaviors, tensile strength, and elongation at break. This is because EPO increased the crosslink density of the NR matrix as dispersing aids.

2.3.3. Polyurethane Elastomer

Polyurethane (PU) elastomers have a great deal of structure and properties because they can be synthesized with a wide variety of materials. For this reason, it has been used in many applications such as shoes, tires, construction, sports, electricity, etc. For the synthesis of PU elastomers, diisocyanate such as aromatic or aliphatic, macrodiol such as polymeric diol, and small molecule diol or diamine as chain extenders are used to form a copolymer. Macrodiol sequence is a soft-segment and diisocyanate and chain extender sequence are a hard-segment to have the character of elastomer [75–78]. In order to provide mechanical properties such as the elastomer of the PU, the soft segment generally has a glass transition temperature substantially lower than the desired service temperature, and the hard segment has a glass transition temperature or melting temperature much higher than the desired service temperature.

These characteristics of the PU elastomer attract much attention as the matrix of MRE. This is because properties such as tensile strength, stiffness, chemical resistance and friction coefficient of the PU elastomer matrix can be controlled well by changing the type of polyol and diisocyanate according to the application of MRE. Wu et al. [79] manufactured isotropic MREs in an in-situ one-step polycondensation procedure using CI as magnetic particles in the PU base. They improved the phase separation by using surface milling and ball milling of CI particles to increase the dispersibility of CI in the PU elastomer. On the other hand, Hu et al. [80] studied MREs with PU/Si-rubber hybrids and found that they had higher MR effects than pure Si-rubber or PU matrices.

2.3.4. Ethylene/Acrylic Elastomer

Ethylene/acrylic elastomer (AEM) is known to have good low temperature properties, good resistance to compression settings, and combine the deterioration effects of heat, oil, and weather, making it be used in the automobile industries as seal and gasket [81].

Recently, Gao et al. [82] adopted the AEM for an elastomeric matrix of CI-based MRE, also using calcium carbonate as a compatibilizer to study its effects. Due to the characteristics of the positive zeta potential of calcium carbonate, it was suitable as a compatibilizer for the AEM-based MRE. As a result, the addition of calcium carbonate can improve the mechanical strength of MRE as well as the interfacial interaction of CI particles and AEM matrix along with the study of effect of different CI contents on the MRE with AEM matrix. In addition, the interfacial reaction between the CI particles and the matrix was also controlled by different manufacturing methods [83].

Relatively low Mooney viscosity of the AEM is expected such that the CI particles are easy to be dispersed better in the matrix compared to higher Mooney viscosity elastomers such as NR. The MR effect and mechanical properties were compared according to the difference between the content of CI particles and the interfacial reaction with the matrix.

2.3.5. Waste Tire Rubber

Focusing on the environmental and recycling issue, Ubaidillahs et al. [84] reported waste tire rubber based MRE and studied viscoelastic properties of MRE, in which fully vulcanized scrap tire rubber was introduced as the new matrix for the MRE. The successful processing of the scrap tire rubber was undertaken using a high-temperature high-pressure (HTHP) sintering process, converting the inert scrap tires into recycled rubber. They fabricated MRE by running the HPHT sintering process at 25 MPa and 200 °C for an hour. They presented the possibility of making waste tires into highly valued recycled products through the manufacture of MRE from waste material.

2.3.6. Ethylene-Propylene-Diene Monomer Rubber

Ethylene propylene diene monomer (EPDM) rubber is a type of synthetic rubber used in many applications, being widely adopted in the manufacture of cables, hoses, car tire sidewalls, cover strips, wires, belts, shoes and sporting goods, because it possesses good resistance to oxidation, aging and temperature. Additionally, EPDM rubber is easier to handle and cheaper compared to NR. Conversely, EPDM rubber has several disadvantages compared to NR, such as lower tensile strength and fatigue strength. Therefore, carbon black is used a lot as reinforcement for EPDM rubber [85].

Burgaz et al. [86] researched the effects of magnetic particles and carbon black on EPDM rubber based MRE, also focusing on their difference by making MREs with two different micron-sized iron particles of CI powder and bare iron (BI) powder. Through the analysis of Payne effect, they compared the results that CI particles are found to be more compatible with EPDM rubber matrix than BI particles, based on mechanical properties and MR properties according to the contents of CI particles in EPDM rubber based MRE.

Additionally, Plachy et al. [87] fabricated MRE by dispersing CI in EPDM rubber matrix. The degree of porous MRE was controlled through the foaming agent based on azodicarbonamide. Through this, the effect of the porous matrix on the EPDM rubber based MRE was studied. As a result, the porous system complements the high toughness of conventional MREs, which can potentially be utilized in other applications with properties between MR fluids and solid elastomers.

2.3.7. Thermoplastic Elastomer

While most MREs are based on the chemically cross-linked elastomeric matrices, rather easily processable thermoplastic elastomers (TPEs) have been introduced. As desirable alternatives to typical vulcanized rubbers, they can be processed at elevated temperatures as conventional thermoplastics while maintaining their high elasticity. Lu et al. [88] adopted poly(styrene-ethylene-butylene-styrene) (SEBS) triblock copolymer as a TPE to fabricated CI based SEBS MRE and found that the isotropic composite prepared exhibited a larger storage modulus compared to the SEBS matrix at room temperature where the EB phase therein was rubbery while the PS phase was in the glassy state. In contrast, the SEBS composite prepared under the magnetic field at high T contained a chain structure of CI particles. On the other hand, improved thermal stability and mechanical strength of SEBS matrix, not sacrificing the elasticity and toughness was also observed [89].

Furthermore, as a potential injection-molding process for the MRE, polyolefin-based TPEs are also used [90,91]. Cvek et al. [90] fabricated MRE based on propylene-based TPE. Despite decreased MR effect of the MREs by tens of percent per the processing cycle, they expect that the injection-molded TPE-based MREs could offer a new pathway for producing the smart engineering composites owing to the ability to be easily reprocessed.

2.4. Non-Magnetic Fillers

Researches have been conducted to enhance the mechanical and MR characteristics of MREs by adding non-magnetic fillers as additives to MREs. Different types of non-magnetic fillers include reinforcing agents [92–95], plasticizers [96–98], and crosslink agents [99,100] that are related to the mechanical characteristics of MREs, in addition to the important magnetic fillers to increase MR properties. Mechanical reinforcing agents include carbon nanotube (CNT) [92], graphene [93], graphite [94], and carbon black [95]. Li et al. [101] fabricated MRE by adding multi-walled CNTs. The addition of a small amount of CNTs can effectively increase the mechanical properties of MREs. In addition to increasing dynamic stiffness and damping without an applied magnetic field, it also showed a higher field-induced increment. Chen et al. [102] fabricated MREs with different carbon black content and observed the microstructure and mechanical performance of the prepared MREs. As the carbon black is added, the MR effect, damping ratio, and tensile strength of the MRE are all increased.

On the other hand, plasticizers are the most common additive in rubbery materials. Generally, plasticizers are added to the mixing and vulcanization processes of elastomers to increase the flexibility, distensibility, and workability of elastomers. The addition of plasticizer to the MRE helps the magnetic particles to be arranged by the magnetic field strength during the curing of the elastomer, thus increasing the absolute MR effect. Khairi et al. [97] examined the addition of silicone oil as a plasticizer to the SR based MRE. It resulted in an increase of particle alignment and the MR effect. The addition of silicone oil generally lowers the viscosity of the uncured rubber, increasing the alignment of the magnetic particles. Kimura et al. [98] analyzed the properties of MREs by adding a plasticizer such as dioctyl phthalate, dimethyl phthalate, butyl benzyl phthalate and bis(2-methyl octyl) phthalate. As the plasticizer content increases, the G' at 0 mT decreases and the G' at 500 mT increases. This is because the mobility of magnetic particles of MRE is increased by the addition of plasticizer. Fan et al. [103] varied the sulfur content to examine the effect of the crosslink density of the MRE on the damping behaviors. As the crosslinking density of MRE decreases, the movement of magnetic particles increases, and the damping property increases. This confirms that the rearrangement of magnetic particles plays an important role in improving magnetic field-induced change of loss factor, G', and relative MR performance.

While most additives for MREs are normally considered as non-magnetic, recently, addition of magnetic filler additives to the typical CI-based MRE composites has been reported to show a significant effect on the MR properties. Magnetic nano-sized fillers include both soft-magnetic and hard-magnetic nanoparticles. Lee et al. [104] prepared MRE by using micron-sized CI as a main magnetic particle in the NR matrix and adding nano-sized gamma-ferrite as an additive. The gamma-ferrite is a rod-shaped hard magnetic particle, which has been confirmed to be more uniformly aligned with the CI particles of the elastomer added gamma-ferrite than the neat CI-base elastomer due to the morphological characteristics. In addition, strain sweep tests confirmed that MREs with gamma-ferrite added had higher modulus. Kramarenko et al. [105] fabricated MRE by using SR as matrix and CI and NdFeB, a hard-magnetic filler particle, in various sizes and concentrations. MR properties and viscoelastic behavior were measured according to NdFeB size and concentration. The addition of hard magnetic NdFeB caused the non-linear viscoelastic behavior in small strains and caused a large increase in modulus.

3. Characteristics

3.1. Morphological Property

In the MRE, the suspension of magnetic particles and the adhesion between the CI particles and the medium are very important to the properties of the MRE. Morphology of the isotropic and anisotropic MREs is observed through scanning electron microscope (SEM) [97,104]. To observe the morphology of the MRE, the sample is soaked in liquid nitrogen and cut perpendicularly. Through this morphology observation, the adhesion between magnetic particles and matrix and dispersion state of magnetic particles can be observed. Figure 3a is an isotropic MRE prepared without a magnetic field and Figure 3b is SEM image mapped by EDX of anisotropic MRE prepared under magnetic field conditions during curing. Figure 3a shows randomly dispersed carbonyl iron particles in SR. Figure 3b is SEM images of anisotropic MRE sample. Unlike Figure 3a image, the CI particles are aligned in the elastomeric medium, and this confirms that it is anisotropic MREs. These morphological characteristics are deeply associated to the mechanical and MR characteristics of MREs and are important factors for applications.

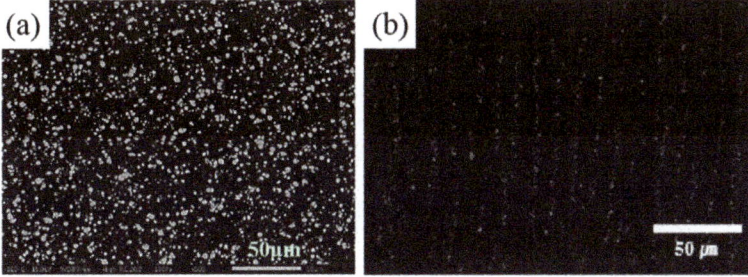

Figure 3. SEM images of the pure CI particles based isotropic MRE (**a**) [97] and anisotropic (**b**) MRE [104].

3.2. Mechanical Characteristics

3.2.1. Tensile Property

In general, as the amount of magnetic particles in the MRE increases, the MR effect increases but the mechanical property decreases [106,107]. The deterioration of these mechanical characteristics is a very important factor in the industrial application of MREs. For this reason, the measurement of the mechanical characteristics of MREs is very important. A good way to measure the mechanical properties of MREs is to measure their tensile strength. Wu et al. [79] conducted a study comparing the properties of surface treated CI and untreated CI on PU-based MREs. Table 1 represents the mechanical characteristics of PU and PU-CI composites with 50%, 60%, and 70% of raw and treated CI. As shown in Table 1, both the tensile strength and the elongation at break decreased with increased CI content. However, when the CI content is 70%, the comparison shows that the MRE made of the surface-treated carbonyl iron has higher tensile strength and elongation at break than the MRE made of the untreated CI. This shows that the surface treatment of CI increases the affinity of the matrix and magnetic particles, resulting in higher mechanical properties.

Table 1. Mechanical characteristics of anisotropic polyurethane (PU) MREs with different CI concentrations [79].

Iron Contents (%)	Tensile Strength/MPa		Elongation at Break/%		Stress at 100% Strain/MPa		Stress at 200% Strain/MPa		Stress at 300% Strain/MPa	
	Raw	Treated	Raw	Treated	Raw	Treated	Raw	Treated	Raw	Treated
0	9.54		466		1.78		2.33		3.26	
50	9.58	12.07	984	971	1.86	1.92	2.03	2.18	2.32	2.65
60	6.93	7.63	1129	1152	1.66	1.79	1.80	1.97	2.04	2.24
70	1.89	2.34	327	417	1.92	2.23	2.00	2.30	1.99	2.27

3.2.2. Payne Effect

Payne [108] reported that the effects of changes due to deformation of the microstructure of the material were due to breakage and reform of weak physical bonds connecting neighboring filler clusters. This analysis has been applied for the elastomeric composites where the filler particles are connected to the elastomer in the microstructural viewpoint. As the strain amplitude increases, the distance between the CI particles increases and the bond between the elastomer and the particle breaks. For this reason, G' decreases as the structure breaks. In other words, if the bond strength of particles and matrix is strong, the change of G' will be small due to the change of strain amplitude. If the bond strength of particles and elastomer is weak, the change of G' will be large due to strain amplitude. In other words, the Payne effect represents the bond strength between the elastomeric medium and the filler particles. Many researchers have evaluated the properties of MREs through the Payne effect of anisotropic and isotropic MREs [109–112]. The Payne effect is measured by the change of G' according to the change of strain amplitude under cyclic loading conditions and calculated using Equation (1),

$$\text{Payne Effect} = \frac{G'_0 - G'_\infty}{G'_0} \times 100 \quad (1)$$

where G'_0 is the G' at an initial strain and G'_∞ is the G' at the infinite strain. Figure 4 is a graph showing the G' of both pure CI based MRE and PGMA coated CI-based MRE as a function of strain [60]. The Payne effect of the CI-based MRE was calculated to be 84%, while that of PGMA coated CI-based MRE was 78%. MRE with PGMA coated CI had a lower Payne effect. This confirms that CI coated with PGMA has a stronger bond strength with the rubber matrix than the CI-based MRE.

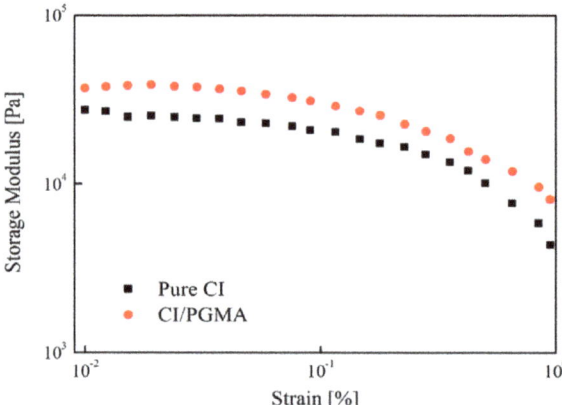

Figure 4. The storage moduli CI/PGMA and pure CI based magnetorheological (MR) elastomers [60].

3.2.3. Damping Loss Factor

The loss factor which is also called damping factor, tan δ = G″/G′, represents the ratio of G′ to loss modulus (G″) and the degree of energy dissipation during the deformation of the material. It is adopted to measure the damping efficiency of MREs [113,114]. Dissipation occurs mainly at the interface between the matrix and the particles, that is, higher interfacial adhesion results in loss factor reduction. Hapipi et al. [115] examined plate-like CI-based MREs (MRE-P) and spherical type CI-based MREs (MRE-S). The field-dependent rheological properties were studied. Figure 5 shows the loss factor according to the frequency of both MRE-P and MRE-S. In all samples, the loss factor tends to increase with an increased frequency, whereas the loss factor of the MRE-P sample is lower than that of MRE-S. Since the primary factor of the loss factor is related to the interface interaction between the particle and the elastomer, the following results indicate that the relative motion between the particles is increased due to the weak interface bonding between the particles and the medium, which results in energy dissipation. The stronger the interfacial interaction between particles and the medium, the smaller the degree of loss factor. As a result, it confirmed that plate-like CI has stronger interfacial bonds with matrix than spherical CI particles.

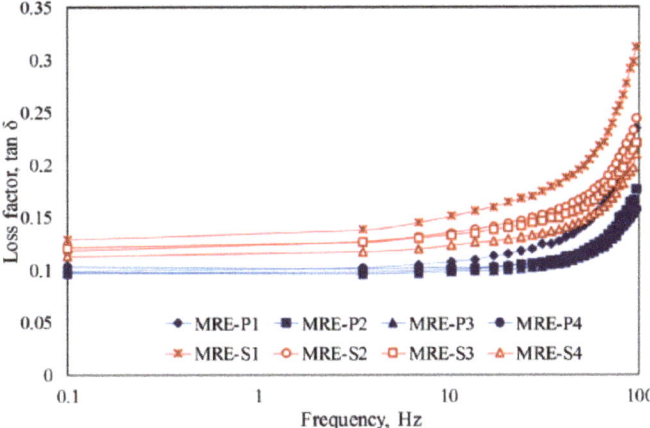

Figure 5. Frequency dependence of loss factor with different curing magnetic fields and particle shapes [115].

3.3. Magnetic Property

The magnetic characteristics of the CI particles of the MREs play an important role in their MR properties [116,117]. Knowing the magnetization information of the magnetic particles provides important information for applying the magnetic particles to the MRE. This magnetization investigation is obtained in the form of hysteresis loops as a graph of magnetization value versus magnetic field strength. Figure 6a shows the magnetization curves of pristine CI particles and particles covered with PGMA on the CI surface [60]. Since CI particles are soft magnetics, they show narrow magnetic hysteresis loops. For pure CI, saturation magnetization was 198 Am^2/kg and CI/PGMA was 153 Am^2/kg, respectively, confirming that the saturation magnetization is reduced by the coating of PGMA. Figure 6b is the magnetization curve of CI particles coated with poly(trimethylsilyloxyethyl methacrylate) (PHEMATMS) [118]. It indicated that the saturation magnetization value decreased as the PHEMATMS coating thickness increased from 11 nm to 22 nm.

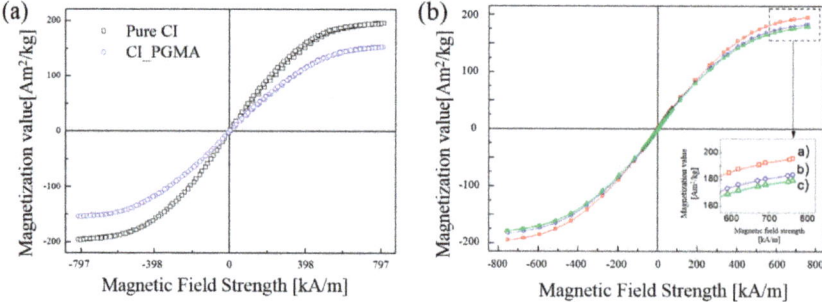

Figure 6. (**a**) VSM curves of pure CI (open dark square) particles and PGMA coated CI (open blue circle) particles [60] and (**b**) magnetization curves of a—bare CI, b—CI-g-PHEMATMS-1 (11 nm), and c—CI-g-PHEMATMS-2 (22 nm) particles obtained via VSM [118].

3.4. Rheological Characteristics

3.4.1. Dynamic Viscoelastic Property

Both amplitude and frequency sweep tests are being conducted to observe the dynamic mechanical behavior of MREs. MREs are viscoelastic materials that generate and store energy during the deformation process. In general, the ability to store energy is associated to the G′, representing the elastic properties of the material. G″ shows the amount of energy dissipation during the deformation process [119]. These G′ and G″ are important parameters for representing viscoelastic properties such as MRE.

Amplitude Sweep Test

The strain amplitude sweep test examines effect of shear strain on the dynamic moduli of the MRE sample [120]. This test allows to define a linear viscoelastic (LVE) region of the MRE. Figure 7a shows the G′ of the MRE sample as a function of strain amplitude [121]. The strain range is measured from 0.01% to 1% and is measured at 3 different magnetic field intensities. As shown in the figure, the low strain region has a plateau region where the G′, called LVE regime, remains constant. As the strain increases, the G′ tends to decrease gradually, and as the strength of the magnetic field increases, the G′ rises.

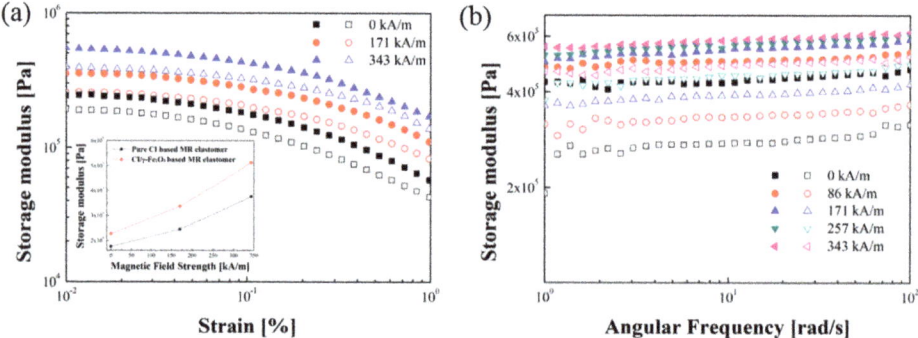

Figure 7. (**a**) Storage Modulus versus strain amplitude sweep and (**b**) storage modulus versus angular frequency sweep (Closed symbol:CI/γ-Fe$_2$O$_3$ based MRE, open symbol: Pure CI based MRE) [121].

Frequency Sweep Test

Figure 7b is a frequency sweep test graph [121] of both G' and G" displayed according to frequency. The constant strain was determined as 0.02% in the LVE region of the strain amplitude sweep test. The measured data range is 1~100 rad/s. As the magnetic field strength increases, the dynamic modulus of the sample increases. At the same magnetic field strength, the G' of CI/γ-Fe$_2$O$_3$ based MRE is higher than that of pure CI based MRE as in the amplitude sweep test.

MR Effect

MR effect is one of the key parameters for characterizing the performance of MREs. The change in the behaviors of the MR composite under the magnetic field is expressed by the absolute MR effect and the relative MR effect [122–125]. The absolute MR effect represents the difference between the shear modulus of the maximum value obtained under the magnetic field and the shear modulus (zero field modulus) when there is no magnetic field as given in Equation (2),

$$\Delta G = G_{max} - G_0 \tag{2}$$

On the other hand, the relative MR effect is the percentage of absolute MR effect and zero field modulus G_0 as follows;

$$\Delta G_r = \frac{\Delta G}{G_0} \times 100\% \tag{3}$$

As well-known in many studies, the absolute and relative MR effects vary with several factors, including oscillation frequency, magnetic field strength, magnetic particle content, particle size, magnetic material, and the MRE matrix. Khairi et al. [97] studied the effects of silicone oil plasticizer on MREs. Figure 8 shows the relative MR effects of isotropic and anisotropic MREs depending on the amount of silicone oil. Both isotropic and anisotropic MREs show increased relative MR effects as silicone oil content increases. The reason is that the maximum value of shear modulus is similar even though the contents of silicone oil are increased, but as the content of silicone oil is increased, the zero-field modulus (G_0) of MRE is lowered, which causes a sharp increase in relative MR effect. As such, the MR effect is considered to be a key parameter in evaluating the performance of the MRE.

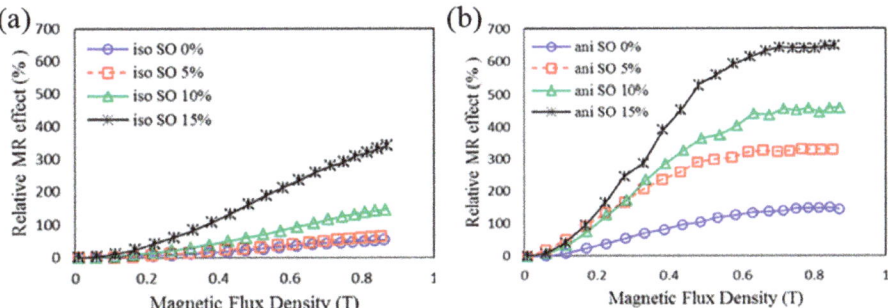

Figure 8. Relative MR effect for (**a**) isotropic MRE and (**b**) anisotropic MRE as a function of magnetic field for different SO contents [97].

3.4.2. Creep and Recovery Characteristics

Creep test is an experiment to observe the change of strain with time when certain stress is applied to a sample. In other words, it is a test about the time-dependent mechanical behavior of a sample. Recovery behavior is defined as the change in strain over time that occurs when a material's stress is abruptly removed. These creep and recovery experiments can be used to infer changes in the microstructure of viscoelastic materials. This helps to understand the properties of the material and the rheological properties of the material. These creep and recovery characteristics play an important role in some engineering applications. In addition, the study of creep and recovery behavior in MREs is very helpful in understanding the deformation mechanism of MR elastomers [126–130].

Bica et al. [131] prepared MREs by varying the volume fraction of the magnetic particles of the PU base to 10%, 20%, and 40%, and then creep and recovery experiments were conducted. As shown in Figure 9, as soon as the constant stress of 30 Pa is induced, the change of strain occurs immediately. In the case of the sample without a magnetic field, as the amount of CI particles increases, the response strain of the sample decreases, and when the magnetic field is input, the MRE without the magnetic field is measured. This shows that an external magnetic field can cause large stiffness of the MRE. Furthermore, Bica et al. [126] fabricated MREs based on silicone rubber with different CI contents. In this study, as a new application of the MR elastomer, they draw their interest in the achievement of electric capacitors whose capacity, in addition to MR responses, is controlled by an outside magnetic field, along with its preparation and viscoelastic characteristics. They measured creep and recovery curves at constant magnetic field (171 kA/m) and under same stress (30 Pa). As a result, as the content of magnetic particles increased, the recovery ratio decreased.

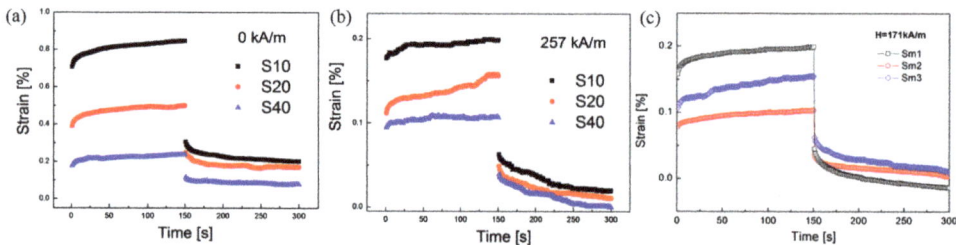

Figure 9. Creep and recovery curves of MREs without (**a**) and with (**b**) the magnetic field (H = 257 kA/m) [131] (**c**) creep and recovery curves of pure elastomer and the MRE samples with the stimuli of magnetic field [126].

4. Applications of MREs

The studies of the controlling both damping and stiffness of MRE has attracted enormous interests in recent years. The field dependence and controllable of mechanical performance of MRE allows to suit for wide engineering applications as follows.

4.1. Vibration Absorber

The various conventional dynamic vibration absorbers (DVAs) are adopted to lower the forced vibration excited at a specific frequency. However, one of drawback of DVAs is the lacking adaptability due to narrow frequency bandwidth. Number of development have been conducted to overcome the disadvantages such as controlling the pressure of air springs [132], altering effective coil number of spring [133], changing the length of threaded flexible rods [134], and minimizing the space between two spring leaves [135]. It is viable to adjust the frequency, the challenges are heavy weight, slow adjusting speed, large dimensions and high energy consumption. Hence one of the high arising solution is MRE. Ginder et al. [136] conducted pioneering work on development on actively controllable MRE vibration absorber. Deng et al. [137] designed series of MRE absorbers working in shear mode, which are able to tune the frequency from 55 to 82 Hz. In addition, Hoang et al. [138,139] fabricated a conceptual ATVA with MRE to minimize vibration inherent for vehicle power train systems.

4.2. Vibration Isolator

The smart MRE isolation system, whose stiffness and damping characteristics differ according to the external environment, has been implanted in several studies to overpower the drawbacks of conventional rubber isolators. Many studies have been researched on MRE isolators. Opie et al. [140]

researched room temperature vulcanized SR to produce a vertically tunable isolator, which increased the resonance peak attenuation and load speed up to 16% and 30%, respectively. Liao et al. [141] fabricated MRE isolator with a voice coil motor to reduce the transfer rate up to 70 %. Li et al. [142,143] reported a lateral laminated MRE bearing for large scale building isolation, which includes multilayer thin MRE sheets attached on to multilayer thin steel plates. The maximum lateral stiffness that can be changed due to different excitation currents is up to 1630%. In addition, Xing et al. [144], similar bearing structure to Li et al. [143], but for seismic mitigation in a bridge superstructure. The resonance frequency has increased from 10 Hz to 20 Hz, which applies to seismic mitigation. The four MRE isolator embedded together, which is used for the bridge monitoring equipment is given in Figure 10b.

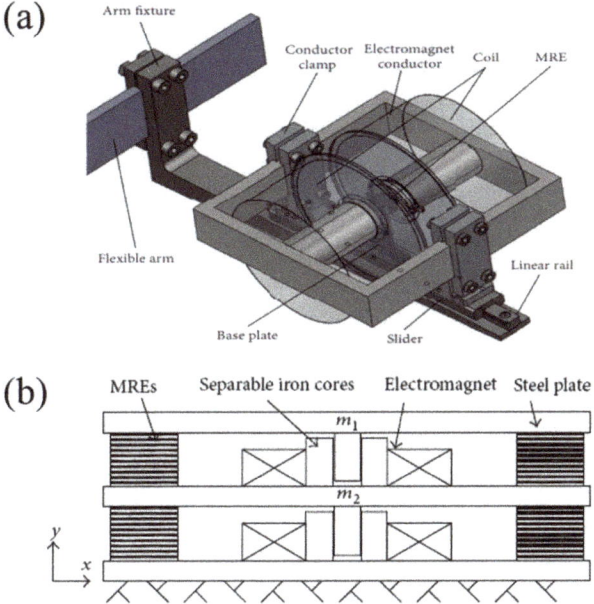

Figure 10. Schematic diagram of (**a**) MRE vibration absorber (Reprinted with permission from [145] and (**b**) MRE isolation system [146].

4.3. Magnetoresistor Sensor

It is well-known that electrical conductivity of MREs can be affected by the intensity of the applied magnetic field allowing them to be used in various applications [94]. However, MRE is not categorized as electric conductors. MRE is capable of operating as active devices of the electric circuit that allows to control output voltage in magnetic field by addition of iron and graphite microparticles. Therefore, one of the applications of MRE could be a magnetoresistor sensor device. Bica [147] reported that MRE is applicable for quadrupolar magnetoresistor and active element of electric circuit. The preparation of electroconductive MRE was conducted by polymerization of SR with addition of silicone oil, CI microparticles, graphite powder, and catalyst. The electrical conductivity of MRE was found to be affected by magnetic field intensity. Thereby, the voltage at the output terminals of the MR sensor device is considerably influenced by the intensity of the transverse magnetic field, for control voltages kept constant. Hence, electrical conductivity is influenced by magnetic field intensity, applied voltage and time [148].

4.4. Electromagnetic Wave Absorber

MREs have been attempted not only in vibration isolators or absorbers but also applicable to electromagnetic wave absorption very recently [149]. The F-Fe/MRE composite was prepared by compression molding and vulcanization using flaky CI particles and SR matrix. The 28 wt.% F-Fe/MRE with 4.3 mm thickness, shows the optimal reflection loss value of −53.3 dB at 4.8 GHz and effective frequency bandwidth (RL < −10 dB) of 6.0 GHz. The combination of magnetic loss, dielectric loss and impedance results matches the multiple reflections and scatterings [149]. In addition, the hybrid MRE is fabricated using silicone oil, rubber, CI particle, graphene nanoparticles and cotton fabric. The dielectric constant and electrical conductivity of hybrid MRE was observed to be enhanced according to the external magnetic field due to relaxation polarization of the SR and interfacial polarization of the graphene [150]. For a potential application on radio-absorbing materials and piezoresistive sensors, Moucka et al. [151] examined relationship between the dielectric response of MREs and their filler clusters' morphology. Regarding the orientation effect of CI particles, it was further reported that the orientation of CI particles strongly enhances the permittivity of the systems, while preserving their permeability, which ultimately manifests itself in enhanced absorption of electromagnetic energy and reduced thickness of radio-absorbers. Thus, radio-absorbers based on anisotropic MREs are characterized by superior EM shielding capability in the microwave frequency range compared to their isotropic analogues [152].

4.5. Other Applications

MRE can be adopted in many other applications. According to Hu et al. [31], the silicone elastomer that consists hard magnetic neodymium-iron-boron (NdFeB) microparticles of 5 µm size was able to act as magneto-elastic soft milimetre-scale robot. This soft robot can move around on surface of liquid, walk and roll on solid surfaces, and crawl in narrow tunnels. The fabrication of small-scale continuum robots with materials of silicone and magnetic particle (NdFeB) enables to navigate through constrained and complex environment for medical applications [153]. In addition, the origami-inspired rapid prototype robot has been fabricated for self-folding, reconfigurable shape, and controllable motility with poly (ethylene glycol) diacrylate and magnetite particles [154].

MRE is also viable for micro-fluid actuation applications. The manipulation of micro-fluid is crucial in biological analysis, optics, chemical synthesis, and information technology. The micro-fluid actuation was achieved using artificial cilia and iron particle with polymer matrix [155]. Another actuated artificial cilia was made using polybutylacrylate and paramagnetic magnetite nano particles to simulate the cilia motion [156]. The MRE has also been used in biomedical application in such a way as active scaffolds for drug and cell delivery. The new active porous scaffold that can be controlled by a magnetic field to transfer different biological agents using alginate hydrogel and magnetite particles [157]. The formation of biofilm on medical device causes serious infections. Hence the engineered tunable active surface topographies with micron-sized pillars was developed using magnetite particles, which was controlled by electromagnetic field [158]. In addition, the MRE could be also applicable in shape-programmable magnetic soft matter, generating new time-varying shapes of soft matter with magnetization profile and actuating field, using ferromagnetic and aluminum particles [159].

Finally, MRE can be used in acoustic metamaterials. The transformation of 3D phase of magnetoactive lattice structures from 2D phase by magnetic field allows to switching signs of constitutive parameters repeatedly. It opens new potential of MRE in acoustic transportation, imaging and refraction [160].

5. Conclusions

In this review, we briefly overviewed fabrication, characterizations, properties, and applications of smart MREs composed of various magnetic particles, elastomers, and additives based on the recent studies. Especially, various polymeric elastomeric materials including waste tire rubber are extensively covered. While the morphology of iso/anisotropic MREs was typically observed with SEM images along with their synthesis and physical characteristics, the mechanical properties of MREs are explained based on tensile strength, Payne effect and loss factor. The MR performance of anisotropic MRE was superior than that of isotropic MRE. Accordingly, the mechanical characteristics of surface coated based MRE was observed stronger than the pure CI particle coated MRE due to the enhanced bonding energy between rubber matrix and CI particle. The MR analysis of MRE under various range of magnetic field is conducted with strain amplitude and frequency sweep. According to the MR property measurements, the G' increases with high magnetic field and at various strain. Hence, the MRE has been drawn a great attention on various applications, such as absorber, isolator, sensor and EMI shielding.

Author Contributions: S.S.K. and K.C. prepared the draft together with the guidance of H.J.C and J.-D.N., H.J.C. initiated and finalized this review paper. S.S.K. and K.C. contributed equally to this work. All authors have read and agreed to the published version of the manuscript.

Funding: This work was supported by National Research Foundation of Korea, grant number 2018R1A4A1025169.

Conflicts of Interest: The authors declare no conflict of interest.

References

1. Chow, W.S.; Mohd Ishak, W.S. Smart polymer nanocomposites: A review. *eXPRESS Polym. Lett.* **2020**, *14*, 416–435. [CrossRef]
2. Zhang, W.L.; Choi, H.J. Graphene oxide based smart fluids. *Soft Matter* **2014**, *10*, 6601–6608. [CrossRef] [PubMed]
3. Olabi, A.G.; Grunwald, A. Design and application of magneto-rheological fluid. *Mater. Des.* **2007**, *28*, 2658–2664. [CrossRef]
4. Carlson, J.D.; Jolly, M.R. MR fluid, foam and elastomer devices. *Mechatronics* **2000**, *10*, 555–569. [CrossRef]
5. Thakur, M.K.; Sarkar, C. Influence of Graphite Flakes on the Strength of Magnetorheological Fluids at high Temperature and Its Rheology. *IEEE Trans. Magn.* **2020**, *56*, 4600210. [CrossRef]
6. Cho, M.S.; Lim, S.T.; Jang, I.B.; Choi, H.J.; Jhon, M.S. Encapsulation of Spherical Iron-Particle With PMMA and Its Magnetorheological Particles. *IEEE Trans. Magn.* **2004**, *40*, 3036–3038. [CrossRef]
7. Chuah, W.H.; Zhang, W.L.; Choi, H.J.; Seo, Y. Magnetorheology of core–shell structured carbonyl iron/polystyrene foam microparticles suspension with enhanced stability. *Macromolecules* **2015**, *48*, 7311–7319. [CrossRef]
8. Jang, W.H.; Kim, J.W.; Choi, H.J.; Jhon, M.S. Synthesis and electrorheology of camphorsulfonic acid doped polyaniline suspensions. *Colloid Polym. Sci.* **2001**, *279*, 823–827. [CrossRef]
9. Seo, Y.P.; Han, S.; Choi, J.; Takahara, A.; Choi, H.J.; Seo, Y. Searching for a Stable High-Performance Magnetorheological Suspension. *Adv. Mater.* **2018**, *30*, 1704769. [CrossRef]
10. Zhang, P.; Lee, K.-H.; Lee, C.-H. Friction behavior of magnetorheological fluids with different material types and magnetic field strength. *Chin. J. Mech. Eng.* **2016**, *29*, 84–90. [CrossRef]
11. Zubieta, M.; Eceolaza, S.; Elejabarrieta, M.; Bou-Ali, M. Magnetorheological fluids: Characterization and modeling of magnetization. *Smart Mater. Struct.* **2009**, *18*, 095019. [CrossRef]
12. Furst, E.M.; Gast, A.P. Micromechanics of magnetorheological suspensions. *Phys. Rev. E* **2000**, *61*, 6732–6740. [CrossRef] [PubMed]
13. Ashtiani, M.; Hashemabadi, S.; Ghaffari, A. A review on the magnetorheological fluid preparation and stabilization. *J. Magn. Magn. Mater.* **2015**, *374*, 716–730. [CrossRef]
14. Ashtiani, M.; Hashemabadi, S. An experimental study on the effect of fatty acid chain length on the magnetorheological fluid stabilization and rheological properties. *Colloids Surf. A Physicochem. Eng. Asp.* **2015**, *469*, 29–35. [CrossRef]
15. Walke, G.; Patil, A.; Rivankar, S. Magnetorheological fluid—A review on characteristics, devices and applications. *J. Emerg. Technol.* **2019**, *6*, 46–50.
16. Kumar, J.S.; Paul, P.S.; Raghunathan, G.; Alex, D.G. A review of challenges and solutions in the preparation and use of magnetorheological fluids. *Int. J. Mech. Mater. Eng.* **2019**, *14*, 13. [CrossRef]

17. Skalski, P.; Kalita, K. Role of magnetorheological fluids and elastomers in today's world. *Acta Mech. Autom.* **2017**, *11*, 267–274. [CrossRef]
18. Bodniewicz, D.; Kaleta, J.; Lewandowski, D. The fabrication and the identification of damping properties of magnetorheological composites for energy dissipation. *Compos. Struct.* **2018**, *189*, 177–183. [CrossRef]
19. Szmidt, T.; Pisarski, D.; Konowrocki, R.; Awietjan, S.; Boczkowska, A. Adaptive Damping of a Double-Beam Structure Based on Magnetorheological Elastomer. *Shock. Vib.* **2019**, *2019*, 8526179. [CrossRef]
20. Lian, C.; Lee, K.H.; Choi, S.B.; Lee, C.H. A study of the magnetic fatigue properties of a magnetorheological elastomer. *J. Intell. Mater. Syst. Struct.* **2019**, *30*, 749–754. [CrossRef]
21. Koo, J.-H.; Dawson, A.; Jung, H.-J. Characterization of actuation properties of magnetorheological elastomers with embedded hard magnetic particles. *J. Intell. Mater. Syst. Struct.* **2012**, *23*, 1049–1054. [CrossRef]
22. Wang, Y.; Hu, Y.; Chen, L.; Gong, X.; Jiang, W.; Zhang, P.; Chen, Z. Effects of rubber/magnetic particle interactions on the performance of magnetorheological elastomers. *Polym. Test.* **2006**, *25*, 262–267. [CrossRef]
23. Shit, S.C.; Shah, P. A review on silicone rubber. *Natl. Acad. Sci. Lett.* **2013**, *36*, 355–365. [CrossRef]
24. Sun, T.L.; Gong, X.L.; Jiang, W.Q.; Li, J.F.; Xu, Z.B.; Li, W.H. Study on the damping properties of magnetorheological elastomers based on cis-polybutadiene rubber. *Polym. Test.* **2008**, *27*, 520–526. [CrossRef]
25. Zhang, W.; Gong, X.L.; Sun, T.L.; Fan, Y.C.; Jiang, W.Q. Effect of Cyclic Deformation on Magnetorheological Elastomers. *Chin. J. Chem. Phys.* **2010**, *23*, 226–230. [CrossRef]
26. Lokander, M.; Stenberg, B. Performance of isotropic magnetorheological rubber materials. *Polym. Test.* **2003**, *22*, 245–251. [CrossRef]
27. Wei, B.; Gong, X.; Jiang, W. Influence of polyurethane properties on mechanical performances of magnetorheological elastomers. *J. Appl. Polym. Sci.* **2009**, *116*, 771–778. [CrossRef]
28. Yang, X.; Shuai, C.G.; Yang, S.L. Magnetorheological Effect of NDI Polyurethane-Based MR Elastomers. *Adv. Mater. Res.* **2013**, *750–752*, 832–835.
29. Sarkar, P.; Bhowmick, A.K. Sustainable rubbers and rubber additives. *J. Appl. Polym. Sci.* **2018**, *135*, 45701. [CrossRef]
30. Yunus, N.A.; Mazlan, S.A.; Ubaidillah; Aziz, S.A.A.; Shilan, S.T.; Wahab, N.A.A. Thermal stability and rheological properties of epoxidized natural rubber-based magnetorheological elastomer. *Int. J. Mol. Sci.* **2019**, *20*, 746. [CrossRef]
31. Hu, W.; Lum, G.Z.; Mastrangeli, M.; Sitti, M. Small-scale soft-bodied robot with multimodal locomotion. *Nature* **2018**, *554*, 81–85. [CrossRef] [PubMed]
32. Lu, H.; Zhang, M.; Yang, Y.; Huang, Q.; Fukuda, T.; Wang, Z.; Shen, Y. A bioinspired multilegged soft millirobot that functions in both dry and wet conditions. *Nat. Commun.* **2018**, *9*, 1–7. [CrossRef]
33. Gong, X.; Zhang, X.; Zhang, P. Fabrication and characterization of isotropic magnetorheological elastomers. *Polym. Test.* **2005**, *24*, 669–676. [CrossRef]
34. Bocian, M.; Kaleta, J.; Lewandowski, D.; Przybylski, M. Test setup for examination of magneto-mechanical properties of magnetorheological elastomers with use of a novel approach. *Arch. Civil. Mech. Eng.* **2016**, *16*, 294–303. [CrossRef]
35. Li, W.; Zhang, X. A study of the magnetorheological effect of bimodal particle based magnetorheological elastomers. *Smart Mater. Struct.* **2010**, *19*, 035002. [CrossRef]
36. Tian, T.; Nakano, M. Fabrication and characterisation of anisotropic magnetorheological elastomer with 45 iron particle alignment at various silicone oil concentrations. *J. Intell. Mater. Syst. Struct.* **2018**, *29*, 151–159. [CrossRef]
37. Sapouna, K.; Xiong, Y.; Shenoi, R. Dynamic mechanical properties of isotropic/anisotropic silicon magnetorheological elastomer composites. *Smart Mater. Struct.* **2017**, *26*, 115010. [CrossRef]
38. Soledad Antonel, P.; Jorge, G.; Perez, O.E.; Butera, A.; Gabriela Leyva, A.; Martín Negri, R. Magnetic and elastic properties of CoFe2O4-polydimethylsiloxane magnetically oriented elastomer nanocomposites. *J. Appl. Phys.* **2011**, *110*, 043920. [CrossRef]
39. Kchit, N.; Bossis, G. Electrical resistivity mechanism in magnetorheological elastomer. *J. Phys. D Appl. Phys.* **2009**, *42*, 105505. [CrossRef]
40. Mordina, B.; Tiwari, R.K.; Setua, D.K.; Sharma, A. Magnetorheology of Polydimethylsiloxane Elastomer/FeCo3 Nanocomposite. *J. Phys. Chem. C* **2014**, *118*, 25684–25703. [CrossRef]
41. Hadi, N.H.N.A.; Ismail, H.; Abdullah, M.K.; Shuib, R.K. Influence of matrix viscosity on the dynamic mechanical performance of magnetorheological elastomers. *J. Appl. Polym. Sci.* **2020**, *137*, 48492. [CrossRef]

42. Abdul Aziz, S.A.; Mazlan, S.A.; Ubaidillah, U.; Shabdin, M.K.; Yunus, N.A.; Nordin, N.A.; Choi, S.B.; Rosnan, R.M. Enhancement of Viscoelastic and Electrical Properties of Magnetorheological Elastomers with Nanosized Ni-Mg Cobalt-Ferrites as Fillers. *Materials* **2019**, *12*, 3531. [CrossRef]
43. Chen, L.; Gong, X.; Li, W. Microstructures and viscoelastic properties of anisotropic magnetorheological elastomers. *Smart Mater. Struct.* **2007**, *16*, 2645–2650. [CrossRef]
44. Sedlačík, M.; Pavlínek, V.; Sáha, P.; Švrčinová, P.; Filip, P.; Stejskal, J. Rheological properties of magnetorheological suspensions based on core–shell structured polyaniline-coated carbonyl iron particles. *Smart Mater. Struct.* **2010**, *19*, 115008. [CrossRef]
45. Li, W.; Du, H.; Chen, G.; Yeo, S.; Guo, N. Nonlinear rheological behavior of magnetorheological fluids: Step-strain experiments. *Smart Mater. Struct.* **2002**, *11*, 209–217. [CrossRef]
46. Li, W.H.; Du, H.; Chen, G.; Yeo, S.H.; Guo, N. Nonlinear viscoelastic properties of MR fluids under large-amplitude-oscillatory-shear. *Rheol. Acta* **2003**, *42*, 280–286. [CrossRef]
47. Li, W.; Zhang, X.; Du, H. Development and simulation evaluation of a magnetorheological elastomer isolator for seat vibration control. *J. Intell. Mater. Syst. Struct.* **2012**, *23*, 1041–1048. [CrossRef]
48. Puente-Córdova, J.G.; Reyes-Melo, M.E.; Palacios-Pineda, L.M.; Martínez-Perales, I.A.; Martínez-Romero, O.; Elías-Zúñiga, A. Fabrication and characterization of isotropic and anisotropic magnetorheological elastomers, based on silicone rubber and carbonyl iron microparticles. *Polymers* **2018**, *10*, 1343. [CrossRef]
49. Kwon, S.H.; Lee, J.H.; Choi, H.J. Magnetic Particle Filled Elastomeric Hybrid Composites and Their Magnetorheological Response. *Materials* **2018**, *11*, 1040. [CrossRef]
50. Lee, J.Y.; Kwon, S.H.; Choi, H.J. Magnetorheological characteristics of carbonyl iron microparticles with different shapes. *Korea-Aust. Rheol. J.* **2019**, *31*, 41–47. [CrossRef]
51. Han, J.K.; Lee, J.Y.; Choi, H.J. Rheological effect of Zn-doped ferrite nanoparticle additive with enhanced magnetism on micro-spherical carbonyl iron based magnetorheological suspension. *Colloids Surf. A Physicochem. Eng. Asp.* **2019**, *571*, 168–173. [CrossRef]
52. Jinaga, R.E.; Jagadeesha, T. Synthesis of Magnetorheological Fluid using Electrolytic Carbonyl Iron Powder for Enhanced Stability. *IOP Conf. Ser. Mater. Sci. Eng.* **2019**, *577*, 012014. [CrossRef]
53. Zhu, W.; Dong, X.; Huang, H.; Qi, M. Iron nanoparticles-based magnetorheological fluids: A balance between MR effect and sedimentation stability. *J. Magn. Magn. Mater.* **2019**, *491*, 165556. [CrossRef]
54. Baranwal, D.; Deshmukh, D.T.S. MR-Fluid Technology and Its Application—A Review. *Int. J. Emerg. Technol. Adv. Eng.* **2012**, *2*, 563–569.
55. Makarova, L.A.; Alekhina, Y.A.; Omelyanchik, A.S.; Peddis, D.; Spiridonov, V.V.; Rodionova, V.V.; Perov, N.S. Magnetorheological foams for multiferroic applications. *J. Magn. Magn. Mater.* **2019**, *485*, 413–418. [CrossRef]
56. Mohamad, N.; Mazlan, S.A.; Ubaidillah; Choi, S.B.; Nordin, M.F.M. The Field-Dependent Rheological Properties of Magnetorheological Grease Based on Carbonyl-Iron-Particles. *Smart Mater. Struct.* **2016**, *25*, 095043. [CrossRef]
57. Fang, F.F.; Liu, Y.D.; Choi, H.J. Fabrication of Carbonyl Iron Embedded Polycarbonate Composite Particles and Magnetorheological Characterization. *IEEE Trans. Magn.* **2009**, *45*, 2507–2510. [CrossRef]
58. An, J.S.; Kwon, S.H.; Choi, H.J.; Jung, J.H.; Kim, Y.G. Modified silane-coated carbonyl iron/natural rubber composite elastomer and its magnetorheological performance. *Compos. Struct.* **2017**, *160*, 1020–1026. [CrossRef]
59. Li, J.; Gong, X.; Zhu, H.; Jiang, W. Influence of particle coating on dynamic mechanical behaviors of magnetorheological elastomers. *Polym. Test.* **2009**, *28*, 331–337. [CrossRef]
60. Kwon, S.H.; An, J.S.; Choi, S.Y.; Chung, K.H.; Choi, H.J. Poly(glycidyl methacrylate) Coated Soft-Magnetic Carbonyl Iron/Silicone Rubber Composite Elastomer and Its Magnetorheology. *Macromol. Res.* **2019**, *27*, 448–453. [CrossRef]
61. Fuchs, A.; Sutrisno, J.; Gordaninejad, F.; Caglar, M.B.; Yanming, L. Surface polymerization of iron particles for magnetorheological elastomers. *J. Appl. Polym. Sci.* **2010**, *117*, 934–942. [CrossRef]
62. Cvek, M.; Mrlík, M.; Ilčíková, M.; Mosnáček, J.; Münster, L.; Pavlínek, V. Synthesis of Silicone Elastomers Containing Silyl-Based Polymer-Grafted Carbonyl Iron Particles: An Efficient Way to Improve Magnetorheological, Damping, and Sensing Performances. *Macromolecules* **2017**, *50*, 2189–2200. [CrossRef]
63. Park, B.J.; Park, C.W.; Yang, S.W.; Kim, H.M.; Choi, H.J. Core-shell typed polymer coated-carbonyl iron suspensions and their magnetorheology. *J. Phys. Conf. Ser.* **2009**, *149*, 012078. [CrossRef]
64. Moksin, N.; Ismail, H.; Abdullah, M.K.; Shuib, R.K. Magnetorheological Elastomer Composites Based on Industrial Waste Nickel Zinc Ferrite and Natural Rubber. *Rubber Chem. Technol.* **2019**, *92*, 749–762. [CrossRef]

65. Borin, D.Y.; Stepanov, G.V.; Odenbach, S. Tuning the tensile modulus of magnetorheological elastomers with magnetically hard powder. *J. Phys. Conf. Ser.* **2013**, *412*, 012040. [CrossRef]
66. Stepanov, G.V.; Chertovich, A.V.; Kramarenko, E.Y. Magnetorheological and deformation properties of magnetically controlled elastomers with hard magnetic filler. *J. Magn. Magn. Mater.* **2012**, *324*, 3448–3451. [CrossRef]
67. Stepanov, G.V.; Borin, D.Y.; Kramarenko, E.Y.; Bogdanov, V.V.; Semerenko, D.A.; Storozhenko, P.A. Magnetoactive elastomer based on magnetically hard filler: Synthesis and study of viscoelastic and damping properties. *Polym. Sci. Ser. A+* **2014**, *56*, 603–613. [CrossRef]
68. Ioppolo, T.; Ötügen, M.V. Magnetorheological polydimethylsiloxane micro-optical resonator. *Opt. Lett.* **2010**, *35*, 2039. [CrossRef]
69. Zhang, X.; Peng, S.; Wen, W.; Li, W. Analysis and fabrication of patterned magnetorheological elastomers. *Smart Mater. Struct.* **2008**, *17*, 45001. [CrossRef]
70. Xu, Z.; Wu, H.; Wang, Q.; Jiang, S.; Yi, L.; Wang, J. Study on movement mechanism of magnetic particles in silicone rubber-based magnetorheological elastomers with viscosity change. *J. Magn. Magn. Mater.* **2020**, *494*, 165793. [CrossRef]
71. Li, W.H.; Nakano, M. Fabrication and characterization of PDMS based magnetorheological elastomers. *Smart Mater. Struct.* **2013**, *22*, 055035. [CrossRef]
72. Bokobza, L. Natural Rubber Nanocomposites: A Review. *Nanomaterials* **2018**, *9*, 12. [CrossRef] [PubMed]
73. Chen, L.; Gong, X.-L.; Jiang, W.-Q.; Yao, J.-J.; Deng, H.-X.; Li, W.-H. Investigation on magnetorheological elastomers based on natural rubber. *J. Mater. Sci.* **2007**, *42*, 5483–5489. [CrossRef]
74. Aziz, S.A.A.; Mazlan, S.A.; Ismail, N.I.N.; Ubaidillah; Choi, S.B.; Nordin, N.A.; Mohamad, N. A comparative assessment of different dispersing aids in enhancing magnetorheological elastomer properties. *Smart Mater. Struct.* **2018**, *27*, 117002. [CrossRef]
75. Xie, F.; Zhang, T.; Bryant, P.; Kurusingal, V.; Colwell, J.M.; Laycock, B. Degradation and stabilization of polyurethane elastomers. *Prog. Polym. Sci.* **2019**, *90*, 211–268. [CrossRef]
76. Zhang, Y.; Fang, F.; Huang, W.; Chen, Y.; Qi, S.; Yu, M. Dynamic Mechanical Hysteresis of Magnetorheological Elastomers Subjected to the Cyclic Loading and Periodic Magnetic Field. *Front. Mater.* **2019**, *6*, 292. [CrossRef]
77. Ju, B.; Tang, R.; Zhang, D.; Yang, B.; Yu, M.; Liao, C.; Yuan, X.; Zhang, L.; Liu, J. Dynamic mechanical properties of magnetorheological elastomers based on polyurethane matrix. *Polym. Compos.* **2016**, *37*, 1587–1595. [CrossRef]
78. Watanabe, M.; Takeda, Y.; Maruyama, T.; Ikeda, J.; Kawai, M.; Mitsumata, T. Chain Structure in a Cross-Linked Polyurethane Magnetic Elastomer under a Magnetic Field. *Int. J. Mol. Sci.* **2019**, *20*, 2879. [CrossRef] [PubMed]
79. Wu, J.; Gong, X.; Chen, L.; Xia, H.; Hu, Z. Preparation and characterization of isotropic polyurethane magnetorheological elastomer through in situ polymerization. *J. Appl. Polym. Sci.* **2009**, *114*, 901–910. [CrossRef]
80. Hu, Y.; Wang, Y.L.; Gong, X.L.; Gong, X.Q.; Zhang, X.Z.; Jiang, W.Q.; Zhang, P.Q.; Chen, Z.Y. New magnetorheological elastomers based on polyurethane/Si-rubber hybrid. *Polym. Test.* **2005**, *24*, 324–329. [CrossRef]
81. Sahoo, B.P.; Naskar, K.; Tripathy, D.K. Conductive carbon black-filled ethylene acrylic elastomer vulcanizates: Physico-mechanical, thermal, and electrical properties. *J. Mater. Sci.* **2011**, *47*, 2421–2433. [CrossRef]
82. Gao, T.; Chen, J.; Hui, Y.; Wang, S.; Choi, H.J.; Chung, K. Role of calcium carbonate as an interfacial compatibilizer in the magneto-rheological elastomers based on ethylene/acrylic elastomer (AEM) and its magneto-induced properties. *Mater. Res. Express* **2019**, *6*, 085320. [CrossRef]
83. Gao, T.; Xie, R.; Chung, K. Microstructure and dynamic mechanical properties of magnetorheological elastomer based on ethylene/acrylic elastomer prepared using different manufacturing methods. *Micro Nano Lett.* **2018**, *13*, 1026–1030. [CrossRef]
84. Ubaidillah; Choi, H.J.; Mazlan, S.A.; Imaduddin, F.; Harjana, H. Fabrication and viscoelastic characteristics of waste tire rubber based magnetorheological elastomer. *Smart Mater. Struct.* **2016**, *25*, 115026. [CrossRef]
85. Kang, I.; Khaleque, M.A.; Yoo, Y.; Yoon, P.J.; Kim, S.-Y.; Lim, K.T. Preparation and properties of ethylene propylene diene rubber/multi walled carbon nanotube composites for strain sensitive materials. *Compos. A Appl. Sci. Manuf.* **2011**, *42*, 623–630. [CrossRef]

86. Burgaz, E.; Goksuzoglu, M. Effects of magnetic particles and carbon black on structure and properties of magnetorheological elastomers. *Polym. Test.* **2020**, *81*, 106233. [CrossRef]
87. Plachy, T.; Kratina, O.; Sedlacik, M. Porous magnetic materials based on EPDM rubber filled with carbonyl iron particles. *Compos. Struct.* **2018**, *192*, 126–130. [CrossRef]
88. Lu, X.; Qiao, X.; Watanabe, H.; Gong, X.; Yang, T.; Li, W.; Sun, K.; Li, M.; Yang, K.; Xie, H.; et al. Mechanical and structural investigation of isotropic and anisotropic thermoplastic magnetorheological elastomer composites based on poly(styrene-b-ethyleneco-butylene-b-styrene) (SEBS). *Rheol. Acta* **2012**, *51*, 37–50. [CrossRef]
89. Qiao, X.; Lu, X.; Gong, X.; Yang, T.; Sun, K.; Chen, X. Effect of carbonyl iron concentration and processing conditions on the structure and properties of the thermoplastic magnetorheological elastomer composites based on poly(styrene-b-ethylene-co-butyleneb-styrene) (SEBS). *Polym. Test.* **2015**, *47*, 51–58. [CrossRef]
90. Cvek, M.; Kracalik, M.; Sedlacik, M.; Mrlik, M.; Sedlarik, V. Reprocessing of injection-molded magnetorheological elastomers based on TPE matrix. *Compos. Part B* **2019**, *172*, 253–261. [CrossRef]
91. Volpe, V.; D'Auria, M.; Sorrentino, L.; Davino, D.; Pantani, R. Injection molding of magneto-sensitive polymer composites. *Mater. Today Commun.* **2018**, *15*, 280–287. [CrossRef]
92. Kumar, V.; Lee, D.J. Mechanical properties and magnetic effect of new magneto-rheological elastomers filled with multi-wall carbon nanotubes and iron particles. *J. Magn. Magn. Mater.* **2019**, *482*, 329–335. [CrossRef]
93. Bica, I.; Anitas, E.M.; Chirigiu, L. Magnetic field intensity effect on plane capacitors based on hybrid magnetorheological elastomers with graphene nanoparticles. *J. Ind. Eng. Chem.* **2017**, *56*, 407–412. [CrossRef]
94. Bica, I. Influence of the transverse magnetic field intensity upon the electric resistance of the magnetorheological elastomer containing graphite microparticles. *Mater. Lett.* **2009**, *63*, 2230–2232. [CrossRef]
95. Nayak, B.; Dwivedy, S.K.; Murthy, K.S.R.K. Fabrication and characterization of magnetorheological elastomer with carbon black. *J. Intell. Mater. Syst. Struct.* **2014**, *26*, 830–839. [CrossRef]
96. Wu, J.; Gong, X.; Fan, Y.; Xia, H. Improving the magnetorheological properties of polyurethane magnetorheological elastomer through plasticization. *J. Appl. Polym. Sci.* **2012**, *123*, 2476–2484. [CrossRef]
97. Khairi, M.H.A.; Fatah, A.Y.A.; Mazlan, S.A.; Ubaidillah, U.; Nordin, N.A.; Ismail, N.I.N.; Choi, S.B.; Aziz, S.A.A. Enhancement of Particle Alignment Using Silicone Oil Plasticizer and Its Effects on the Field-Dependent Properties of Magnetorheological Elastomers. *Int. J. Mol. Sci.* **2019**, *20*, 4085. [CrossRef]
98. Kimura, Y.; Kanauchi, S.; Kawai, M.; Mitsumata, T.; Tamesue, S.; Yamauchi, T. Effect of Plasticizer on the Magnetoelastic Behavior for Magnetic Polyurethane Elastomers. *Chem. Lett.* **2015**, *44*, 177–178. [CrossRef]
99. Stoll, A.; Mayer, M.; Monkman, G.J.; Shamonin, M. Evaluation of highly compliant magneto-active elastomers with colossal magnetorheological response. *J. Appl. Polym. Sci.* **2014**, *131*, 39793. [CrossRef]
100. Hu, T.; Xuan, S.; Ding, L.; Gong, X. Stretchable and magneto-sensitive strain sensor based on silver nanowire-polyurethane sponge enhanced magnetorheological elastomer. *Mater. Des.* **2018**, *156*, 528–537. [CrossRef]
101. Li, R.; Sun, L.Z. Dynamic mechanical behavior of magnetorheological nanocomposites filled with carbon nanotubes. *Appl. Phys. Lett.* **2011**, *99*, 131912. [CrossRef]
102. Chen, L.; Gong, X.L.; Li, W.H. Effect of carbon black on the mechanical performances of magnetorheological elastomers. *Polym. Test.* **2008**, *27*, 340–345. [CrossRef]
103. Fan, Y.; Gong, X.; Xuan, S.; Qin, L.; Li, X. Effect of Cross-Link Density of the Matrix on the Damping Properties of Magnetorheological Elastomers. *Ind. Eng. Chem. Res.* **2012**, *52*, 771–778. [CrossRef]
104. Lee, C.J.; Kwon, S.H.; Choi, H.J.; Chung, K.H.; Jung, J.H. Enhanced magnetorheological performance of carbonyl iron/natural rubber composite elastomer with gamma-ferrite additive. *Colloid Polym. Sci.* **2018**, *296*, 1609–1613. [CrossRef]
105. Kramarenko, E.Y.; Chertovich, A.V.; Stepanov, G.V.; Semisalova, A.S.; Makarova, L.A.; Perov, N.S.; Khokhlov, A.R. Magnetic and viscoelastic response of elastomers with hard magnetic filler. *Smart Mater. Struct.* **2015**, *24*, 035002. [CrossRef]
106. Wen, Q.; Shen, L.; Li, J.; Xuan, S.; Li, Z.; Fan, X.; Li, B.; Gong, X. Temperature dependent magneto-mechanical properties of magnetorheological elastomers. *J. Magn. Magn. Mater.* **2020**, *497*, 165998. [CrossRef]
107. Perales-Martínez, I.A.; Palacios-Pineda, L.M.; Lozano-Sánchez, L.M.; Martínez-Romero, O.; Puente-Cordova, J.G.; Elías-Zúñiga, A. Enhancement of a magnetorheological PDMS elastomer with carbonyl iron particles. *Polym. Test.* **2017**, *57*, 78–86. [CrossRef]

108. Payne, A.R. Effect of dispersion on the dynamic properties of filler-loaded rubbers. *J. Appl. Polym. Sci.* **1965**, *9*, 2273–2284. [CrossRef]
109. Ramier, J.; Gauthier, C.; Chazeau, L.; Stelandre, L.; Guy, L. Payne effect in silica-filled styrene–butadiene rubber: Influence of surface treatment. *J. Polym. Sci. Part B Polym. Phys.* **2007**, *45*, 286–298. [CrossRef]
110. Lion, A.; Kardelky, C. The Payne effect in finite viscoelasticity. *Int. J. Plast.* **2004**, *20*, 1313–1345. [CrossRef]
111. Sorokin, V.V.; Ecker, E.; Stepanov, G.V.; Shamonin, M.; Monkman, G.J.; Kramarenko, E.Y.; Khokhlov, A.R. Experimental study of the magnetic field enhanced Payne effect in magnetorheological elastomers. *Soft Matter* **2014**, *10*, 8765–8776. [CrossRef] [PubMed]
112. Suo, S.; Xu, Z.; Li, W.; Gan, Y. Improved Mathematical Model for Analysis of the Payne Effect of Magnetorheological Elastomers. *J. Aerosp. Eng.* **2018**, *31*, 04018046. [CrossRef]
113. Fan, Y.; Gong, X.; Xuan, S.; Zhang, W.; Zheng, J.; Jiang, W. Interfacial friction damping properties in magnetorheological elastomers. *Smart Mater. Struct.* **2011**, *20*, 035007. [CrossRef]
114. Tong, Y.; Dong, X.; Qi, M. Payne effect and damping properties of flower-like cobalt particles-based magnetorheological elastomers. *Compos. Commun.* **2019**, *15*, 120–128. [CrossRef]
115. Hapipi, N.; Aziz, S.A.A.; Mazlan, S.A.; Ubaidillah; Choi, S.B.; Mohamad, N.; Khairi, M.H.A.; Fatah, A.Y.A. The field-dependent rheological properties of plate-like carbonyl iron particle-based magnetorheological elastomers. *Results Phys.* **2019**, *12*, 2146–2154. [CrossRef]
116. Bunoiu, M.; Bica, I. Magnetorheological elastomer based on silicone rubber, carbonyl iron and Rochelle salt: Effects of alternating electric and static magnetic fields intensities. *J. Ind. Eng. Chem.* **2016**, *37*, 312–318. [CrossRef]
117. Diguet, G.; Sebald, G.; Nakano, M.; Lallart, M.; Cavaillé, J.-Y. Magnetic particle chains embedded in elastic polymer matrix under pure transverse shear and energy conversion. *J. Magn. Magn. Mater.* **2019**, *481*, 39–49. [CrossRef]
118. Cvek, M.; Mrlik, M.; Sevcik, J.; Sedlacik, M. Tailoring Performance, Damping, and Surface Properties of Magnetorheological Elastomers via Particle-Grafting Technology. *Polymers* **2018**, *10*, 1411. [CrossRef]
119. Hadi, N.H.N.A.; Shuib, R.K. Effect of Plasticizer on Microstructure and Dynamic Mechanical Performance of Anisotropic Magnetorheological Elastomers. *IOP Conf. Ser. Mater. Sci. Eng.* **2019**, 548.
120. Tian, T.F.; Zhang, X.Z.; Li, W.H.; Alici, G.; Ding, J. Study of PDMS based magnetorheological elastomers. *J. Phys. Conf. Ser.* **2013**, *412*, 012038. [CrossRef]
121. Kwon, S.H.; Lee, C.J.; Choi, H.J.; Chung, K.H.; Jung, J.H. Viscoelastic and mechanical behaviors of magneto-rheological carbonyl iron/natural rubber composites with magnetic iron oxide nanoparticle. *Smart Mater. Struct.* **2019**, *28*, 045012. [CrossRef]
122. Esmaeilnezhad, E.; Choi, H.J.; Schaffie, M.; Gholizadeh, M.; Ranjbar, M. Polymer coated magnetite-based magnetorheological fluid and its potential clean procedure applications to oil production. *J. Clean. Prod.* **2018**, *171*, 45–56. [CrossRef]
123. Wang, Y.; Zhang, X.; Chung, K.; Liu, C.; Choi, S.-B.; Choi, H.J. Formation of core–shell structured complex microparticles during fabrication of magnetorheological elastomers and their magnetorheological behavior. *Smart Mater. Struct.* **2016**, *25*, 115028. [CrossRef]
124. Dong, X.; Ma, N.; Qi, M.; Li, J.; Chen, R.; Ou, J. The pressure-dependent MR effect of magnetorheological elastomers. *Smart Mater. Struct.* **2012**, *21*, 075014. [CrossRef]
125. Jiang, W.-Q.; Yao, J.-J.; Gong, X.-L.; Chen, L. Enhancement in Magnetorheological Effect of Magnetorheological Elastomers by Surface Modification of Iron Particles. *Chin. J. Chem. Phys.* **2008**, *21*, 87–92. [CrossRef]
126. Bica, I.; Liu, Y.D.; Choi, H.J. Magnetic field intensity effect on plane electric capacitor characteristics and viscoelasticity of magnetorheological elastomer. *Colloid Polym. Sci.* **2012**, *290*, 1115–1122. [CrossRef]
127. Li, W.; Zhou, Y.; Tian, T.; Alici, G. Creep and recovery behaviors of magnetorheological elastomers. *Front. Mech. Eng.* **2010**, *5*, 341–346. [CrossRef]
128. Xu, Y.; Gong, X.; Xuan, S.; Li, X.; Qin, L.; Jiang, W. Creep and recovery behaviors of magnetorheological plastomer and its magnetic-dependent properties. *Soft Matter* **2012**, *8*, 8483–8492. [CrossRef]
129. Yu, W.X.; Yang, S.L.; Yang, X. Creep Property of Magnetorheological Elastomers of Seism Isolator for Building. *Appl. Mech. Mater.* **2013**, *357–360*, 1291–1294.
130. Qi, S.; Yu, M.; Fu, J.; Zhu, M. Stress relaxation behavior of magnetorheological elastomer: Experimental and modeling study. *J. Intell. Mater. Syst. Struct.* **2017**, *29*, 205–213. [CrossRef]

131. Bica, I.; Anitas, E.M.; Averis, L.M.E.; Kwon, S.H.; Choi, H.J. Magnetostrictive and viscoelastic characteristics of polyurethane-based magnetorheological elastomer. *J. Ind. Eng. Chem.* **2019**, *73*, 128–133. [CrossRef]
132. Brennan, M. Vibration control using a tunable vibration neutralizer. *Proc. Inst. Mech. Eng. Part C J. Mech. Eng. Sci.* **1997**, *211*, 91–108. [CrossRef]
133. Franchek, M.; Ryan, M.; Bernhard, R. Adaptive passive vibration control. *J. Sound Vib.* **1996**, *189*, 565–585. [CrossRef]
134. Hill, S.G.; Snyder, S.D. Design of an adaptive vibration absorber to reduce electrical transformer structural vibration. *J. Vib. Acoust.* **2002**, *124*, 606–611. [CrossRef]
135. Walsh, P.L.; Lamancusa, J. A variable stiffness vibration absorber for minimization of transient vibrations. *J. Sound Vib.* **1992**, *158*, 195–211. [CrossRef]
136. Ginder, J.M.; Schlotter, W.F.; Nichols, M.E. Magnetorheological elastomers in tunable vibration absorbers. *Smart Mater. Struct.* **2001**, *4331*, 103–110.
137. Deng, H.-X.; Gong, X.-L.; Wang, L.-H. Development of an adaptive tuned vibration absorber with magnetorheological elastomer. *Smart Mater. Struct.* **2006**, *15*, N111–N116. [CrossRef]
138. Hoang, N.; Zhang, N.; Du, H. A dynamic absorber with a soft magnetorheological elastomer for powertrain vibration suppression. *Smart Mater. Struct.* **2009**, *18*, 074009. [CrossRef]
139. Hoang, N.; Zhang, N.; Li, W.; Du, H. Application of a magnetorheological elastomer to develop a torsional dynamic absorber for vibration reduction of powertrain. In Proceedings of the 12th International Conference, Philadelpia, PA, USA, 16–20 August 2010; pp. 179–185.
140. Opie, S.; Yim, W. Design and control of a real-time variable modulus vibration isolator. *J. Intell. Mater. Syst. Struct.* **2011**, *22*, 113–125. [CrossRef]
141. Liao, G.; Gong, X.; Xuan, S.; Kang, C.; Zong, L. Development of a real-time tunable stiffness and damping vibration isolator based on magnetorheological elastomer. *J. Intell. Mater. Syst. Struct.* **2012**, *23*, 25–33. [CrossRef]
142. Li, Y.; Li, J.; Tian, T.; Li, W. A highly adjustable magnetorheological elastomer base isolator for applications of real-time adaptive control. *Smart Mater. Struct.* **2013**, *22*, 095020. [CrossRef]
143. Li, Y.; Li, J.; Li, W.; Samali, B. Development and characterization of a magnetorheological elastomer based adaptive seismic isolator. *Smart Mater. Struct.* **2013**, *22*, 035005. [CrossRef]
144. Xing, Z.-W.; Yu, M.; Fu, J.; Wang, Y.; Zhao, L.-J. A laminated magnetorheological elastomer bearing prototype for seismic mitigation of bridge superstructures. *J. Intell. Mater. Syst. Struct.* **2015**, *26*, 1818–1825. [CrossRef]
145. Bian, Y.; Liang, X.; Gao, Z. Vibration Reduction for a Flexible Arm Using Magnetorheological Elastomer Vibration Absorber. *Shock Vib.* **2018**, *2018*, 9723538. [CrossRef]
146. Li, R.; Du, C.; Guo, F.; Yu, G.; Lin, X. Performance of Variable Negative Stiffness MRE Vibration Isolation System. *Adv. Mater. Sci. Eng.* **2015**, *2015*, 837657. [CrossRef]
147. Bica, I. Magnetoresistor sensor with magnetorheological elastomers. *J. Ind. Eng. Chem.* **2011**, *17*, 83–89. [CrossRef]
148. Bica, I. Influence of the magnetic field on the electric conductivity of magnetorheological elastomers. *J. Ind. Eng. Chem.* **2010**, *16*, 359–363. [CrossRef]
149. Zheng, J.; He, X.; Li, Y.; Zhao, B.; Ye, F.; Gao, C.; Li, M.; Li, X.; E, S. Viscoelastic and magnetically aligned flaky Fe-based magnetorheological elastomer film for wide-bandwidth electromagnetic wave absorption. *Ind. Eng. Chem. Res.* **2020**, *59*, 3425–3437. [CrossRef]
150. Bica, I.; Bunoiu, O.M. Magnetorheological Hybrid Elastomers Based on Silicone Rubber and Magnetorheological Suspensions with Graphene Nanoparticles: Effects of the Magnetic Field on the Relative Dielectric Permittivity and Electric Conductivity. *Int. J. Mol. Sci.* **2019**, *20*, 4201. [CrossRef]
151. Moucka, R.; Sedlacik, M.; Cvek, M. Dielectric properties of magnetorheological elastomers with different microstructure. *Appl. Phys. Lett.* **2018**, *112*, 122901. [CrossRef]
152. Cvek, M.; Moucka, R.; Sedlacik, M.; Babayan, V.; Pavlinek, V. Enhancement of radio-absorbing properties and thermal conductivity of polysiloxane based magnetorheological elastomers by the alignment of filler particles. *Smart Mater. Struct.* **2017**, *26*, 095005. [CrossRef]
153. Kim, Y.; Parada, G.A.; Liu, S.; Zhao, X. Ferromagnetic soft continuum robots. *Sci. Robot.* **2019**, *4*, 7329. [CrossRef]
154. Huang, H.W.; Sakar, M.S.; Petruska, A.J.; Pane, S.; Nelson, B.J. Soft micromachines with programmable motility and morphology. *Nat. Commun.* **2016**, *7*, 12263. [CrossRef] [PubMed]

155. Fahrni, F.; Prins, M.W.; van Ijzendoorn, L.J. Micro-fluidic actuation using magnetic artificial cilia. *Lab Chip.* **2009**, *9*, 3413–3421. [CrossRef]
156. Khaderi, S.N.; Craus, C.B.; Hussong, J.; Schorr, N.; Belardi, J.; Westerweel, J.; Prucker, O.; Ruhe, J.; den Toonder, J.M.; Onck, P.R. Magnetically-actuated artificial cilia for microfluidic propulsion. *Lab Chip.* **2011**, *11*, 2002–2010. [CrossRef]
157. Zhao, X.; Kim, J.; Cezar, C.A.; Huebsch, N.; Lee, K.; Bouhadir, K.; Mooney, D.J. Active scaffolds for on-demand drug and cell delivery. *Proc. Natl. Acad. Sci. USA* **2011**, *108*, 67–72. [CrossRef] [PubMed]
158. Gu, H.; Lee, S.W.; Carnicelli, J.; Zhang, T.; Ren, D. Magnetically driven active topography for long-term biofilm control. *Nat. Commun.* **2020**, *11*, 2211. [CrossRef]
159. Lum, G.Z.; Ye, Z.; Dong, X.; Marvi, H.; Erin, O.; Hu, W.; Sitti, M. Shape-programmable magnetic soft matter. *Proc. Natl. Acad. Sci. USA* **2016**, *113*, E6007–E6015. [CrossRef]
160. Yu, K.; Fang, N.X.; Huang, G.; Wang, Q. Magnetoactive Acoustic Metamaterials. *Adv. Mater.* **2018**, *30*, 1706348. [CrossRef]

Publisher's Note: MDPI stays neutral with regard to jurisdictional claims in published maps and institutional affiliations.

© 2020 by the authors. Licensee MDPI, Basel, Switzerland. This article is an open access article distributed under the terms and conditions of the Creative Commons Attribution (CC BY) license (http://creativecommons.org/licenses/by/4.0/).

Review

Systematic Review on the Effects, Roles and Methods of Magnetic Particle Coatings in Magnetorheological Materials

Siti Khumaira Mohd Jamari [1], Nur Azmah Nordin [1,*], Ubaidillah [2,*], Siti Aishah Abdul Aziz [1], Nurhazimah Nazmi [1] and Saiful Amri Mazlan [1]

1. Malaysia-Japan International Institute of Technology, Universiti Teknologi Malaysia, Jalan Sultan Yahya Petra, Kampung Datuk Keramat, Kuala Lumpur 54100, Malaysia; khumaira_jamari@yahoo.com (S.K.M.J.); aishah118@gmail.com (S.A.A.A.); nurhazimah@utm.my (N.N.); amri.kl@utm.my (S.A.M.)
2. Mechanical Engineering Department, Faculty of Engineering, Universitas Sebelas Maret, Jalan Ir. Sutami 36A, Kentingan, Surakarta 57126, Indonesia
* Correspondence: nurazmah.nordin@utm.my (N.A.N.); ubaidillah_ft@staff.uns.ac.id (U.)

Received: 25 October 2020; Accepted: 23 November 2020; Published: 24 November 2020

Abstract: Magnetorheological (MR) material is a type of magneto-sensitive smart materials which consists of magnetizable particles dispersed in a carrier medium. Throughout the years, coating on the surface of the magnetic particles has been developed by researchers to enhance the performance of MR materials, which include the improvement of sedimentation stability, enhancement of the interaction between the particles and matrix mediums, and improving rheological properties as well as providing extra protection against oxidative environments. There are a few coating methods that have been employed to graft the coating layer on the surface of the magnetic particles, such as atomic transfer radical polymerization (ATRP), chemical oxidative polymerization, and dispersion polymerization. This paper investigates the role of particle coating in MR materials with the effects gained from grafting the magnetic particles. This paper also discusses the coating methods employed in some of the works that have been established by researchers in the particle coating of MR materials.

Keywords: magnetorheological; smart materials; coating; particle coating

1. Introduction

Magnetorheological (MR) material has captured global interests in the smart materials industry and research, due to its alterable functionalities which correspond with the applied magnetic field. The rheological properties of the materials can be varied accordingly with the application of such external stimuli, therefore MR material is considered to be one of the most promising smart materials which can be applied in engine mounts [1], brake systems [2–4] and vibration absorbers and dampers [5–7], as well as in sensors [8–10]. These applications have excellent benefits in industries that involve transportation [11,12], seismic prevention [13], soft robotics [14], and prosthetic legs [15].

The first MR material was developed by Rabinow [16] in 1948 where a "magnetic fluid clutch" was designed, with a type of magnetizable particles that were mixed in a fluid carrier medium. Under the influence of an external magnetic field, the magnetic particles had been "mutually attracted" and the material "seemingly solidifies", as mentioned by Rabinow [16]. The fluid was then known as magnetorheological fluid, or MRF, with the most commonly used fluid carriers being silicone oil [4,17–21], synthetic hydrocarbon [22–25], and mineral oil [26,27]. The mechanism of MRF is as illustrated in Figure 1, where during the absence of an external magnetic field (off-state condition), the material behaves like a Newtonian-fluid, however when the external magnetic field is applied upon the material (on-state condition), the micron-sized magnetizable particles in the fluid align to the direction of the applied field. This causes a restriction of the flow of the carrier fluid, thus changing the behavior of the material to become stiffer and more solid-like. This reversible behavior change can occur almost instantaneously. Due to the chain-like structure of the aligned magnetic particles, yield stress is needed to break the particle chains to allow the flow of the fluid, which means that the higher the applied magnetic field to the material, the stronger the particle chain, and thus the higher the force needed to overcome the yield point where the particle chain begins to deform.

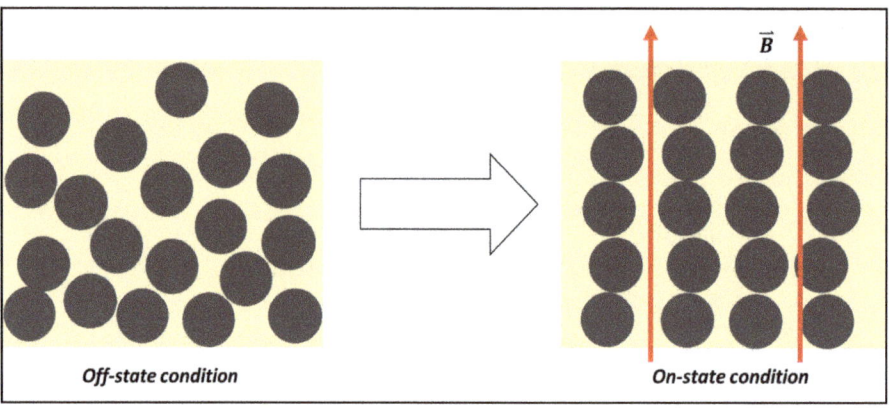

Figure 1. The mechanism of magnetorheological fluid (MRF) prior and upon the application of external magnetic field, $\vec{B}$.

MR grease and MR gel were developed later with a much higher viscosity fluid than MRF to minimize the serious sedimentation problem which occurred in MRF, by reducing the density mismatch between the magnetic particles and the carrier medium. MR grease/gel also has a unique property where it possesses solid gel properties at room temperature, but acts like a fluid at a higher temperature.

Meanwhile, magnetorheological elastomer (MRE) is the solid analogue of MRF, where the fluid is replaced by a non-magnetic elastomer matrix as its carrier medium. By using an elastomeric polymer, such as silicone rubber [28–32], natural rubber [33–35] and polyurethane [5,36,37] as its medium, the sedimentation issue in MRF is completely tackled because the magnetic particles are locked within the matrices. It also offers an alternative to the flowy MRF; MRE does not need any vessel to contain it because it is already in a solid form, and therefore there are no leakage or sealing problems in MRE. The particles in MRE are fixed in the network of the elastic polymer, therefore the particles cannot form chain-like structures like MRF when subjected to an external magnetic field, but can only be polarized and mutually attracted to one another, causing a change in the stiffness of the material. The curing process of the material can be done either with or without the presence of an external magnetic field. During the presence of the field, the magnetic particles are aligned in the direction of the magnetic field, causing a columnar particle structure after the sample has cured. This process is called anisotropic curing. Meanwhile, in the isotropic curing method, the magnetic particles are cured and dispersed in no orderly pattern in the MRE due to the absence of a magnetic field. The damping and the storage modulus, as well as loss modulus of these MREs, can be controlled under the influence of various magnitudes of external magnetic fields.

Other less-reviewed MR materials are MR foam and MR plastomer, where the MR foam uses a spongy, absorbent carrier medium such as polyurethane [38–40] as its carrier medium. The application of MR foam is also unique; it has been reported that MR foam has a potential application in acoustic absorption [41]. On the other hand, MR plastomer is the latest developed MR material with a carrier medium, such as polypropylene glycol and toluene diisocyanate mixture [42,43], polyvinyl alcohol [44,45], and thermoplastic polyurethane [46], which offers a plasticine-like material that has solid-state between MR gel and MRE. Due to the lack of studies around these forms of MR materials, to the best of our knowledge, there is almost no research involving the incorporation of particle coatings into MR foam and MR plastomer, and therefore in this review, we only focus on the impregnation of coated magnetic particles in the other three MR materials.

2. Problems in MR Materials

The advancement of MR materials from one type to another is mainly due to the limitation of one another. For example, because the magnetic particles are much denser than the carrier fluid, sedimentation occurs in MRF during the off-state condition which can lead to another problem—the aggregation of the particles. This aggregation issue may reduce the efficiency of MRF respective to the MR effect that might be reduced over time. Therefore, MR gel/grease was developed to control the particle–medium density mismatch so that it can have a better redispersibility and thus reduce the sedimentation problem. However, because the matrix-based material is highly viscous, MR gel/grease exhibits a high yield stress at the initial state as well as a lower MR effect compared to MRF.

Meanwhile, MRE and MR foam were developed to eliminate the severe sedimentation issue in MRF by embedding the magnetic particles in the elastic and porous matrices, respectively. The leakage and sealant problems in MRF and MR gel/grease has also been overcome by these solid MR materials. However, by locking the magnetic particles in the solid-based matrices, the particles are restricted to move freely and thus inappropriately respond to the applied external magnetic field in the way the particles in MRF do. This affects rheological changes during the off-on state condition in MRE and MR foam, where the MR effect drops several magnitudes in comparison to that of MRF.

Subsequently, the development of MR plastomer may have minimized this MR effect issue, because MR plastomer is reported to have higher MR effect compared to MRE and MR foam, however lower than MRF due to the MR plastomer itself, which exists in the state between both MR materials and possess a high viscosity as well. In addition, the solvent used in MR plastomer may dry up and suffer from a desiccation problem, and therefore it is not suitable for long-term use [44].

Additionally, because the magnetic particles used in MR materials are made up of metal elements such as pure iron and cobalt, oxidation may also take place, especially for long-term use of applications that could further add problems to MR materials. Due to the huge potential that all respective groups of MR materials could offer, progressive research is continuously being undertaken to improve the internal structure of the materials for enhanced properties. Therefore, a particle coating of the magnetic particles in the MR materials has been proposed as one of the ways to tackle the aforementioned problems.

3. Particle Coating in MR Materials

The advantages of coatings in general are known globally; they can provide extra protection for the substrates that is coated as well as enhance the performance of the materials. The coating application on the surface of magnetic particles has emerged in MR studies, to reduce or minimize the issues that have hindered the full capability of MR materials. This is achieved by grafting a layer of polymer onto the surface of the magnetic particles that later will be incorporated in the MR materials. Figure 2 shows an example of a simple coating onto magnetic particles by grafting (3-Aminopropyl) triethoxysilane (APTES) after the surface of the particle has been pre-treated with hydroxyl moieties. On the other hand, Figure 3 shows an example of a micrograph from scanning electron microscopy (SEM) from a successful grafting of coatings onto the magnetic particles as demonstrated by Sutrisno et al. (2013) [47]. It has been proven that particle coating has great benefits to MR materials in terms of improving the sedimentation stability and re-dispersion of the magnetic particles in MRF, protection against wearing and friction, oxidation protection, and enhancement of rheological properties of the MR materials.

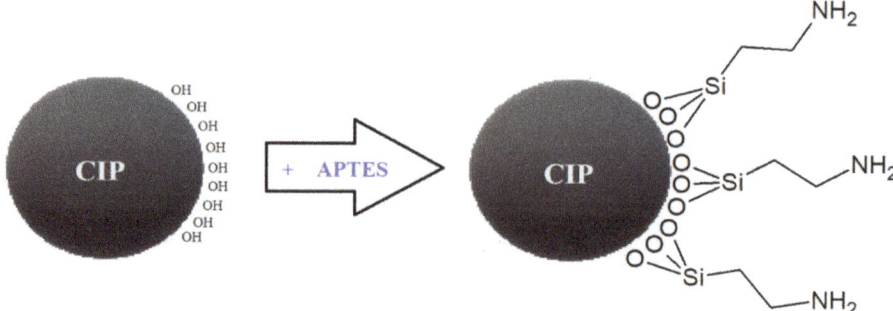

Figure 2. Grafting of (3-Aminopropyl)triethoxysilane (APTES) onto CIP upon activating the surface of the particle with hydroxyl moieties.

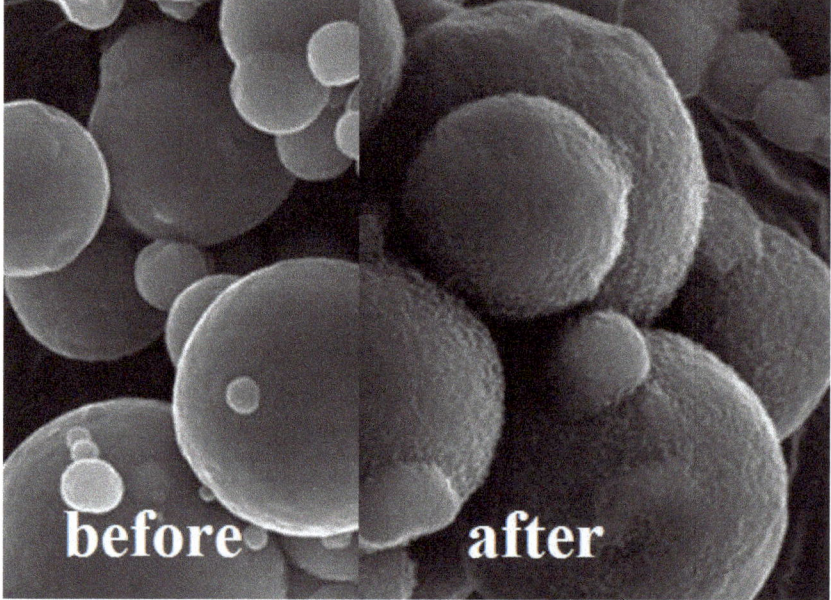

Figure 3. Micrograph of before and after a successful grafting of coating on the magnetic particles (Figure has been modified with reference to [47]).

3.1. Sedimentation Stability

In a study conducted by Hu et al. (2006) [48], it was found that poly(butyl acrylate)-grafted carbonyl iron particles (CIP) showed a significantly lower settling volume of the magnetic particles in commercial MRF. They also discovered that the sedimentation constant of coated CIP was much lower than that of commercial MRF—the higher percentages of grafted-poly(butyl acrylate) exhibited the lowest sedimentation constant. The team explained that these results are due to the coating layer which acts as a stabilizing layer that can sterically prevent coagulation. Meanwhile, Park et al. (2009) [27] demonstrated that by coating the CIP with poly(methyl methacrylate) (PMMA), the sedimentation problem that occurred in their lubricating oil-based MRF improved due to the lower density of the coated magnetic particles compared to uncoated ones. The re-dispersion of the particles after they were subjected to an external magnetic field was also improved, due to enhanced surface properties such as steric repulsion and electrostatic repulsion of the PMMA coating. A further investigation on the sedimentation stability property of coated magnetic particles in MRF has been conducted by Quan et al. (2014) [49], where a Turbiscan instrument was used to compare the transmission percentage of polystyrene-coated CIP MRF and common silicone oil MRF. At the beginning of the experiment, both MRFs showed zero transmission which represented multiple incident lights scattered through the suspension which could not be transmitted due to many particles that were suspended in both MRFs. However, after 400 min, uncoated CIP MRF exhibited a higher transmission percentage compared to polystyrene-coated CIP MRF, indicating the higher sedimentation stability of the coated sample. The Turbiscan was also used by Tae et al. (2017) [50] to compare the sedimentation ratio of polyaniline (PANI)-coated CIP and pure CIP MRFs by subtracting the transmission ratio from 100% value. In this work, it was found that MRF with PANI-coated CIP exhibited a slower sedimentation speed compared to that of uncoated CIP during the whole 900 min of dispersion period. The team stated that this was due to the thick PANI shell that encapsulated the CIP, causing the density of the particles to reduce significantly. Furthermore, they also claimed that the slightly non-spherical and rough surface of the coated particles also contributed to the improved dispersion stability of the MRF. Moreover, it was also revealed that the time response for coated particles in MRF is faster than that of the uncoated ones, as reported by Nguyen et al. (2014) [4]. In this work, the researchers grafted the CIP with silica to be incorporated in MRF for the application of an MR brake, and it was found that the MRF with silica-coated CIP exhibited a higher settling time of 0.450 s compared to 0.342 s for the same MRF with uncoated CIP.

3.2. Tribology Properties

Another significant benefit resulting from particle coating in MRF is related to the unique application of MRF in polishing devices and honing process, as reported by several works [51–54]. Due to the nature of these works, the magnetic particles used in MRF inevitably experience friction against each other and with the contacting surfaces, which will result in wearing problems of the particles. Furthermore, the use of MRF in dampers and isolators can also cause serious wearing problems to the vessel that contains it [26,55]. Therefore, introducing a coating layer on the surface of the magnetic particles could reduce the fretting problem. It was proven by Bombard and de Vicente (2012) [56], when they compared different grades of commercial CIP in MRF, that it acts as a boundary lubricant between polytetrafluoroethylene (PTFE) and a stainless steel tribopair, which was achieved under a pure sliding contact condition. It was found that OS grade CIP possessed the best friction-reduction behavior compared to the five other grades (OM, OX, HS, HSI and HQ), and even when compared to OM grade that is the same size. This is due to the presence of an amorphous silica coating on the surface of the OS CIP that reduces the friction coefficient of the material and indirectly produces the smallest worn scar diameter in comparison to the other grades. Meanwhile, Zhang et al. (2018) [18] subjected two types of silicone oil-based MRFs to a reciprocating friction and wear test; one was filled with polystyrene-coated CIP and the other one with uncoated CIP. Using a surface profilometer, it was found that the width and the depth of the marks produced by MRF with polystyrene-coated CIP were much narrower and smaller compared to that of uncoated CIP during both off-state and on-state conditions, while the surface of the disc that was coated with CIP MRF was analytically and appeared optically smoother than its counterpart after the wear test. They also discovered that the coefficient of friction of the CIP-coated MRF was lower than that of the uncoated-CIP MRF, especially during the on-state condition. These outcomes undisputedly showed that MRF with polystyrene-coated CIP has better wear and frictional properties than MRF with bare CIP.

In other work conducted by Lee et al. (2015) [57], the CIP was coated with PMMA to be applied in their own MR polishing fluid system. They discovered that by using PMMA-coated CIP in their MRF, the material removal depth of the BK7 glass workpiece was lower while exhibiting lower surface roughness than that of the uncoated one with an optimum rotating wheel speed of 1884 mm/s. This superior surface property of the coated CIP exhibited similar pattern outcomes discovered by the same research team of Lee et al. (2017) [54], but this time, the CIP were coated with a biopolymer, xanthan gum. On the other hand, a study by Hong et al. (2018) [58] also showed the decrease in surface roughness of BK7 glass when the CIP used in their MR polishing fluid was coated with silica, especially when an abrasive was introduced into the MRF. These end results can definitely be utilized in ultra-precision applications in optics and micro parts industries. Another notable outcome achieved by coating the magnetic particles for MR polishing fluid application has been expressed by Salzman et al. (2016) [51], where the CIP was coated with zirconia to be applied as a polishing material on a chemical vapor deposited with zinc sulfide substrate (CVD ZnS). One of the few limitations that hinder MRF from being commercially available for polishing applications is the pebble emergence on the surface of the substrate after the polishing work has been done. The researchers have successfully minimized this problem by using zirconia-coated CIP in their MR polishing fluid in a slightly acidic condition of pH4.

Although there is an increasing number of works which investigate the potential of MRE as a polishing material [59,60], to the best of our knowledge, there is no work that involves coated magnetic particles in MRE for polishing application.

3.3. Oxidation

MR materials contain magnetizable particles that inevitably consist of metal elements such as iron, therefore oxidation of the micron-sized particles undoubtedly can occur; this can happen in wet conditions (with the presence of water molecules) or in dry conditions (without the presence of water molecules, which is commonly known as thermo-oxidation).

It has been proven that oxidation phenomena can present negative impacts on the performance of the MR materials, as stated by Lokander et al. (2004) [61]. In their work, the oxidation of the natural rubber of MRE increases as the amount of iron particles increases, due to the oxidation of each individual particle. Furthermore, the high concentration of oxygen elements on the surface of the particles, as well as iron elements of the particles, induces iron ions which are some of the root problems of the oxidation issue of MRE, because the particles are directly in contact with the elastomer matrix. In the meantime, Ulicny et al. (2007) [62] validated the adverse effect of oxidation on MRF after the material was tested for durability performance in an MRF fan clutch. It was found that after 540 h of durability tests, the fan clutch torque capacity reduced gradually as the maximum fan speed experienced dropped by 15%. This is said to be due to the increasing oxidation of the iron particles in conjunction with the fact that iron oxides have much lower magnetization compared to elemental iron. The effects of corrosion in MR materials have also been discussed in detail by Plachy et al. (2018) [17] and Burhannuddin et al. (2020) [63] in separate works around MRF and MRE, respectively. Furthermore, rougher surfaces of oxidized particles may lead to irregular polishing performance as well as reducing the lifetime of the vessels that contain the MR materials, because rougher surfaces can cause unnecessary friction between the particles and the wall of the vessels after a long operational period [57,58,64]. These problems can be minimized by developing coatings that act as a barrier to the oxidation species. It has been proven by Shafrir et al. (2009) [65] in an accelerated corrosion resistance test to determine the oxidation resistance of their zirconia-coated CIP. This test was done by stirring the coated and uncoated CIP in different batches with pH 4.4 acetic acid aqueous solution. After exposure to the accelerated acidic condition for 22 days, the zirconia-coated CIP did not produce any visible goethite (FeOOH) corrosion product. More evidence to the oxidation resistant coating for magnetic particles in MR materials was demonstrated by Cvek et al. (2015) [20], where the team grafted poly(glycidyl methacrylate) (PGMA) onto the CIP to improve the chemical stability of the particles in acidic condition. After exposing the coated and uncoated CIP to 0.05 M HCl in two different containers, the pH values of PGMA-coated CIP remained almost the same throughout the 90 h of acidic exposure, indicating the CIP corrosion protection provided by the PGMA coating. Meanwhile, the performance of coated and uncoated iron particles in MRE before and after an oxidation test has been displayed by Behrooz et al. (2015) [66]. In their study, the poly(tetrafluoropropyl methacrylate)-coated particles and uncoated particles were embedded into silicone rubber MRE. Both samples underwent an accelerated aging test to induce oxidation of the MREs. It was shown that the MRE with coated particles had the least reduction in effective modulus, and the coated particles were able to preserve the stiffness of the MRE under an oxidative environment.

On the other hand, coated magnetic particles in MR materials also exhibited positive outcomes in terms of thermo-oxidation stability. For instance, a facile coating of silicone oil on the CIP by simply immersing the particles in the dimethyl-silicone oil prior being incorporated into MRE, has been proven to enhance the thermal stability as well as the formation of the cross-linked structure of the MRE [67].

3.4. Electromagnetic Shielding

One of the most interesting applications of coated magnetic particles incorporated into MR materials is the ability to shield the materials from electromagnetic waves, which can be used in civil and military fields. This can be achieved when core-shell structure particles, which consist of magnetizable particles coated with electrically conducted shell, are able to absorb unfavorable electromagnetic signals and wave pollution that is subjected to the particles [68]. It has been demonstrated that when tetraethylorthosilicate (TEOS) was grafted onto CIP, the MRE that contained these particles exhibited a higher absorbing ability of ultra-high frequencies of 700 MHz to 1.6 GHz, which is the common operation range of communication and information transmission systems. Through this finding, it was claimed that this was the first time that MRE was reported to have the potential to have sufficient electromagnetic shielding [68].

Furthermore, besides the aforementioned advantages of particle coating in MR materials, there are also other benefits that were obtained indirectly through the coating of the magnetic particles into the MR materials' medium carrier. For instance, in the application of MREs, it was found that the particle's coating has the capability to promote the interfacial interaction between the magnetic particles and elastomer matrices [69], and this would improve inter-bonding between the particles and the matrix. From the perspective of rheological characterization, it was found that due to the enhancement of the particle–matrix interaction in MRE credited to the grafted coating, the mobility of the particles was enhanced compared to the uncoated particles that were locked within the matrix. This particle mobility reinforcement led to a significant increase in damping property and MR effect of the MRE [70].

From these studies, it can be concluded that particle coating in MR materials can deliver more than one purpose when the coated magnetic particles are dispersed in the medium carrier of the materials.

4. Coating Methods

Coating methods employed in particle grafting in MR mainly involve the polymerization of the main monomer on the surface of the particles. This includes Atomic Transfer Radical Polymerization (ATRP), chemical oxidative polymerization, dispersion polymerization and the sol–gel method.

4.1. Atom Transfer Radical Polymerization (ATRP)

ATRP is a type of "living"/controlled polymerization that was first discovered by Wang and Matyjaszewski in 1995 as the expansion of the earlier polymerization method, Atom Transfer Radical Addition (ATRA) [71]. However, compared to ATRA, ATRP requires a reactivation of the first formed alkyl halide-unsaturated monomer added, and further reaction of the irregularly formed radical with propagated monomer units. Known for its ease in preparation, ATRP can precisely control the molecular weight of the produced polymer and thus presents the ability to control the thickness of the grafted coat as well as other sequences such as functionalities and architecture [72,73].

One of the advantages that ATRP has in comparison to other controlled polymerization is that it is catalytic, therefore it can be used for a large number of monomers. A very broad range of molar mass chemical substances containing activated (pseudo) halogen atoms can be used as the initiator. ATRP also allows the replacement of terminal halogens with more useful functional groups to be performed rather easily [74]. The polymerization proceeds with irreversible termination and transfer reactions, therefore the polymers obtained can have a predetermined molecular weight and narrow molecular weight distribution [71]. Furthermore, ATRP can also be conducted in mild conditions (i.e., at room temperature) with a high yield [48].

The ATRP process is mainly composed of a monomer, organic halide initiator/retarder, and a catalyst that consists of transition metal species and solvents. Monomers used in ATRP are structural unit compounds that can withstand propagating radicals such as styrenes [47,75–77], (meth)acrylates [20,70,78–80], acrylates [48,81–83] acrylonitrile [84,85] and acrylamides [86,87].

The reaction in ATRP is catalyzed by transition metal complexes that determine the equilibrium constant between the active and dormant species. This means that the catalyst used determines the polymerization rate. However, this equilibrium constant must not be too small, as the polymerization might be inhibited or slowed down, or it could lead to a wide distribution of chain lengths if the constant is too large. Therefore, it is said that the catalyst is the most important component in ATRP because it determines the position of the atom transfer equilibrium and the dynamics of exchange between the dormant and active species [72]. Cuprous salts such as copper (I) bromide (CuBr) and copper (II) bromide (CuBr2) are the most commonly used catalysts in MR ATRP. Although there are a few studies that used copper chloride as the catalyst, the former is used more extensively due to higher reduction potentials compared to the latter, which can be attributed to a stronger Cu–Br bond in comparison to the Cu–Cl bond. These catalysts are complex, with aliphatic amine, imine or pyridine based ligand such as sparteine [47,48,76], bipyridine [79,88,89] or N,N,N′,N″,N″-pentamethyldiethylenetriamine (PMDETA) [70,81,82,87,90,91] which improve its solubility in a polymerization mixture and fine-tune its catalytic efficiency.

The initiator or dormant propagating chain end, which is usually an alkyl halide (R–X), is included in this type of controlled radical polymerization, where one or more atoms or end groups can be transferred radically. Its role is to react with the monomer to form an intermediate compound. Apart from the monomer conversion, the average molecular weight (Mn) of the synthesized polymers is also dependent on the initial concentration ratio of monomer (M) to the initiator. The halide group (X) must be able to selectively migrate between the growing chain and the transition metal complex rapidly. Therefore, the molecular weight control is the best when bromine or chlorine is used [72]. Commonly used initiators are ethyl α-bromoisobutyrate (EBiB) [20,70,78,81,82,89], α-bromoisobutyryl bromide (BiBB) [70,87,92], methyl 2-bromopropionate [48,75,89] and phenylethyl bromide [75,88]. Additionally, it is noted that there could be more than one initiator used in a single polymerization of ATRP. The number of polymer chains grown for this type of polymerization depends on the type of initiator, because the structure of the initiator influences the architecture of the polymer, as shown in Figure 4.

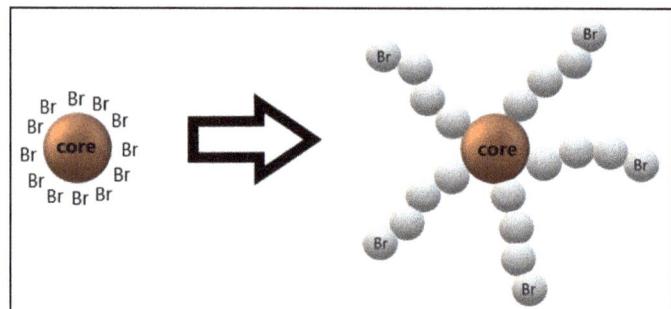

Figure 4. The structure of the bromide initiator influences the star-like architecture of synthesized polymer.

The solvent is the final component of the ATRP coating method, which is also important in this polymerization, especially when the obtained polymer is insoluble in its monomer such as polyacrylonitrile. ATRP can be conducted in bulk or in solution with the usage of non-polar solvents, such as toluene, xylene and anisole (methoxybenzene) or polar solvents such as dimethylformamide

(DMF) [77,81,84,86], acetonitrile [75,81,85] and dimethyl sulfoxide (DMSO) [79,81,84] which are some of the most commonly used solvents in ATRP. However, the type of solvent used must be taken into consideration to minimize the chain transfer to solvent. "Catalyst poisoning" by solvent, which refers to partial or total deactivation of the catalyst due to improper use of solvent must be avoided or at least reduced to ensure efficient ATRP polymerization [93]. Horn and Matyjaszewski (2013) [89] have highlighted the effects of the type of solvent on the activation rate constant with 14 different solvents (anisole, acetonitrile, acetone, butanone, DMSO, DMF, dimethylacetamide, formamide, N-methylpyrrolidone, propylene carbonate, methanol, ethanol, 2-propanol and 2,2,2-trifluoroethanol) which were used to facilitate the polymerization of CuBr/1,1,4,7,10,10-hexamethyltriethylenetetramine (HMTETA) with EBiBB. It was found that the higher the polarity of the solvent, the higher the activation step, while the lower the deactivation step, resulted in a higher the ATRP polymerization constant.

ATRP is one of the most employed coating methods in particle grafting for MR materials. This include two separated works by Cvek et al. (2015) [20,92] where poly(glycidyl methacrylate) (PGMA) was grafted on CIP for the use in MRF. By determining the relative molecular weight using gel permeation chromatography (GPC), relatively narrow polydispersity in both samples were obtained. This indicates that the ATRP process was well controlled and can prevent sedimentation in their MRF samples. Moreover, there was a slight improvement of sedimentation stability in samples which consist of a higher monomer ratio, thus the authors claimed that the length of the polymer chains does not play significant role as expected [92]. This CIP–PGMA sample also exhibited excellent anti-acid corrosion where the coating almost covered the CIP core. However, a small production of hydrogen gas was observed, which indicated that the (3-aminopropyl)triethoxysilane (APTES) base as not fully compact, thus Cl from hydrochloric acid was still able to react with the core [20].

Meanwhile, the grafting of poly(butyl acrylate) was done by Hu et al. (2006) [48] for MRF samples under two separate conditions: a) different butyl acrylate ratio, and b) different reaction time. It was found that the settling was reduced by increasing the coated polymer fraction. The dispersibility of the particles and the sedimentation rate were also greatly improved with the increasing polymer coating fraction. In work done by Sutrisno et al. (2013) [47], poly(2-fluorostyrene) was grafted onto the iron particles for the applications of MRF, while Fuchs et al. (2010) [76] grafted poly(fluorostyrene) onto the iron particles for the applications of MRE. From these works, due to the presence of fluorine, the degradation rate of the coating was low, which implies that it has excellent anti thermo-oxidation property. This has been proven by characterizing the coated samples using differential scanning calorimetry (DSC) and thermogravimetric analysis (TGA). In two separate studies by Cvek et al. (2017) [70] and (2018) [80], the CIP was grafted with poly(trimethylsilyloxyethyl methacrylate) chains to be fabricated in a poly(dimethyl siloxane) (PDMS) MRE matrices. The density of the coated CIP decreased by more than 5% with substantial enhancement of its thermo-oxidation stability. The polymer coating also exhibited extremely stable acidic oxidation, which proved that the grafted layer was uniform without any defects and thus provided an excellent protection against acidic oxidation. Table 1 shows the summary of the ATRP coating method that has been utilized for particle coating in MR materials to date.

With a wide variety of monomers and initiators that can be used in ATRP, there are a lot of options that can be employed to coat the magnetic particles used in MR materials and thus enhance the performance of the materials.

Table 1. Summary of atomic transfer radical polymerization (ATRP) coating method employed by some works.

Monomers	Initiators	Catalysts	Solvents	Remarks	Ref.
Glycidyl methacrylate	BiBB EBiB	CuBr PMDETA (L)	THF Anisole Acetone Ethanol	APTES was used as a coupling agent during surface treatment. CIP was pre-treated with HCl.	[20,92]
Butyl acrylate	CTCS	CuBr CuBr2 Sparteine (L)	THF Toluene		[48]
Fluorinated styrene	CTCS	CuBr CuBr2 Sparteine (L)	Toluene Octyl pyrrolidone		[47,76]
2-hydroxyethyl methacrylate Chlorotrimethyl-silane	BiBB EBiB	CuBr PMDETA (L)	Dichloro-methane Anisole	APTES used as a linker of BiBB Et3N used to trap hydrogen chloride	[70,80]

4.2. Chemical Oxidative Polymerization

Chemical oxidative polymerization (sometimes called chemical oxidation polymerization) is a detachment process of two hydrogen atoms from monomer molecules to form a covalent bond of a polymer [94]. It is regarded as one of the cleanest and lowest loading methods in poly-condensation or condensation polymerization. Commonly used monomers are aromatic amines, aromatic hydrocarbons, thiophenols, phenols and heterocycles, which have high oxidation tendency due to the presence of electron donor substituents. The oxidation of these monomers is performed under the influence of an oxidizing agent (chemically) or by applied potential (electrochemically). Thus, the polymer growth can be initiated due to the generated cation or cation radical sites of the monomer [95]. Therefore, the main components of this polymerization are the electron-donor monomer, the dopant and the oxidant catalyst (initiator).

A conducting polymer is the most common polymer synthesized from chemical oxidative polymerization. It is used as a corrosion inhibitor [96,97], as electromagnetic wave shielding and an absorbance material [98,99], as well as in other electronic devices and semiconductors [94]. The most researched conducting polymer synthesized via chemical oxidative polymerization is polyaniline. It has been widely adopted in conducting polymer-based composites due to its controllable dielectric loss ability, ease of synthesis, and chemical and environmental stability [97–100]. Due to its unique interaction with strong acids as its charge stabilizing agent, polyaniline has three main states: pernigraniline, emeraldine and leucoemeraldine, with each of these states existing in a protonated or deprotonated condition. This means that this polymer can exist in at least six different degrees of oxidation and protonation states [94]. Like other conducting polymers, the properties of polyaniline depend strongly on the doping level, protonation level, ion size of dopant and the water content. The emeraldine base form of polyaniline is an electrical insulator consisting of two amines followed by two imines. It can be converted into a conducting form by two different doping processes, which are protonic acid doping and oxidative doping. In the former, the emeraldine base corresponds to the protonation of imines in which there is no electron exchange, while in the latter, emeraldine salt is obtained from leucoemeraldine through electron exchanges. These reversible mechanisms are caused by the presence of –NH groups in the polymer backbone, whose protonation and deprotonation will bring about a change in the electrical conductivity and the color of the polymer [94].

Lin et al. (2017) [100] experimented the polymerization of polyaniline with various polymerization temperature of −7 °C to 60 °C and found that the formation mechanism as well as the structural and electrical properties of polyaniline emeraldine salt synthesized in strong acidic environment are sensitively affected by the temperature. At high polymerization temperature, the development of nanofibers of aniline is limited, thus the granular emeraldine salt samples possess a weaker hardness, less crystallinity, lower molecular weight and a smaller dispersity compared to those synthesized at lower temperatures. Conventional chemical oxidative polymerization of anilines is carried out in an acidic medium (the dopant) and initiated by an oxidant under an ice bath condition. The most commonly used oxidant is ammonium persulfate (also known as ammonium peroxydisulfate, APS), while some works have mentioned the employment of sodium persulfate [101] and potassium persulfate [97].

Some works have reported the associate polyaniline coating in MR application via chemical oxidative polymerization. For instance, Fan et al. (2013) [99] synthesized CIP/polyaniline composites with APS as the oxidant and p-toluenesulfonic acid (p-TSA) as the dopant with different doped acid mole ratio of p-TSA to aniline (0.005:1; 0.05:1; and 0.2:1 ratios). Obvious core-shell structure of polyaniline on CIP was observed using SEM where the coating layer becomes smoother as the doped acid mole ratio of p-TSA to aniline is higher. It was also reported that by observing the permeability and permittivity behavior, it was noted that the impedance matching between the material and the free space had been increased, which showed the capability of the material to absorb microwave radiation. Although in this work no MR application was mentioned, there is still a possibility that this could be implemented in MR materials because the CIP/polyaniline composite is reported to possess excellent microwave absorbing properties.

Meanwhile, Yu et al. (2017) [69] attempted to improve the interfacial interactions between CIP and polyurethane/epoxy elastomer matrix using polyaniline as an interfacial coating, by making use of the presence of amines and imines in the polymer backbone. The attempt was successful, because the polyaniline-coated CIP and the matrix interfacial interactions were improved due to the covalent bond between these components. The contact angles of the polyaniline-coated CIP displayed an angle 24.3° higher, which suggests an excellent hydrophobic property, while there were obvious increments of storage modulus as well as reduced loss modulus compared to uncoated CIP in the elastomer matrix. In the meantime, a study by Tae et al. (2017) [50] showed the importance of surface treatment of the magnetic particles upon coating with a conducting polymer via chemical oxidative polymerization. In this work, the authors demonstrated the attachment of hydroxyl groups on the surface of CIP using p-TSA monohydrate prior to encapsulating the particles with polyaniline to improve the chemical affinity between polyaniline and the CIP. As a result, the polyaniline-coated CIP had better thermal stabilities, while the sedimentation ratio was greatly improved.

In 2011, a study conducted by He et al. (2011) [98] showed that PANI-coated CIP and PANI-coated iron (II, III) oxide (Fe_3O_4) particles that were synthesized through chemical oxidation polymerization exhibited good microwave absorbing properties. By plotting Cole–Cole semicircle plots, it was claimed that PANI attributed to the dielectric relaxation process of both coated particles. Furthermore, when both PANI-coated particles were mixed to form a composite of PANI/CIP/Fe_3O_4, an appropriate electromagnetic impedance match between these particles was insisted to be the factor that contributes to the enhanced electromagnetic wave absorption of the material. Although the authors did not claim this composite to be used in MR materials, it still has the potential to be applied in this smart material, especially for MRE for electromagnetic wave shielding applications.

To the best of our knowledge, only polyaniline has been associated as the conducting polymer coating for MR materials via chemical oxidative polymerization. However, other conductive polymers that have been reported to be synthesized using this polymerization steps for other applications include polyphenylenediamines, polytoluidine, polypyyrole, polyaminopyridine, polyaminonaphthalene, polyaminoquinoline, polymethylquinoline and polyphenylenediamine, which may open possibilities for ventures into a new type of conductive polymers as the coating layer on the surface of the magnetic particles in MR materials. Some studies of coatings developed via chemical oxidation polymerization in a core-shell structure have been listed below in Table 2 for the potential employment for particle coating application of MR materials. Table 2 includes the type of developed polymers, the type of substrates that were coated onto them, and the findings from the mentioned studies. However, extensive studies must be done to ensure the compatibility of the polymers to be polymerized onto magnetic particles before being incorporated into MR materials, because most of the substrates used are non-magnetic, except [102].

Table 2. Some studies that employed chemical oxidation polymerization outside of the magnetorheological (MR) field for coating applications.

Polymerized Polymers	Substrates	Findings	Ref.
Polypyrrole	Sulfur nano-sphere	For rechargeable lithium/sulfur batteries application - Minimize loss of active materials during cycling	[103]
Polypyrrole	Glass beads	For water treatment application - increase adsorption of humic acid - high positive zeta potentials for wide range of pH values	[104]
Poly(1,5-diaminoanthra-quinone)	Carbon cloth	For supercapacitor application - extraordinary cycling stability - ultrahigh capacitance retention (159%) after 20k cycles	[105]
Polythiophene	Nano-silicon	For lithium batteries application - good cycling stability	[106]
Polyrhodanine	γ-Fe_2O_3 nano-particles	For antibacterial application - reduction in *Escherichia coli* and *Staphylococcus aureus* colonies (90–99.9%) after contact for 60 min	[102]

4.3. Dispersion Polymerization

Dispersion polymerization as defined by IUPAC (2011) is a precipitation-type polymerization in which monomer(s), initiator(s) and colloidal stabilizer(s) are dissolved in a solvent forming an initially homogenous system that produces polymers and results in the formation of polymer particles [107]. This process usually yields polymer particles in colloidal dimensions. It attracts researchers worldwide due to its ability to form monodisperse particles in a single batch process [108]. One of the unique characteristics of dispersion polymerization is that the solvent used as the reaction medium must be one that has compatibility with the monomers used but incompatible with the polymers that are formed. Therefore, the composition of polymers that are polymerized using this method are usually made from common monomers such as styrenes, methacrylates and vinyls, with a good selection of solvents and initiators.

In the MR field, some works that implemented dispersion polymerization method as the coating method includes the work by Park et al. (2009) [27], where the team encapsulated CIP with PMMA to be used in the MRF. In this work, after the CIP were pre-treated with methacrylic acid (MAA), the particles were dispersed in methanol solution that contained poly(vinyl pyrrolidone) (PVP) that acts as a stabilizer. Then the methyl methacrylate (MMA) monomer was dissolved in the reactor system together with 2,2-azobisisobutyronitrile (AIBN) that acts as radical initiator, as well as ethylene glycol dimethacrylate (EDGMA) that acts as the cross-linking agent to produce cross-linked PMMA during this polymerization steps. A reasonably smooth coating with some surface irregularities was produced, while the sedimentation stability of the MRF was enhanced. A similar work of grafting a PMMA coating onto CIP via dispersion polymerization has also been accomplished by Lee et al. (2015) [57], but with further characterization on the performance of the coated CIP in MRF for the application of optical polishing. It was reported that the sedimentation ratio and anti-corrosion property of the coated material had increased, although the yield shear stress of the coated material in a magnetic field was lower than that of uncoated material.

A core-shell structured polystyrene coated on CIP has been demonstrated by Quan et al. (2014) [49] using dispersion polymerization too. In this work, the researchers also used MAA, PVP and AIBN as the surface pre-treatment agent, stabilizer and initiator, respectively. From the SEM results, it was observed that the coating formed on the CIP surface was not uniform, with granule-like substances that were seen scattered unevenly on the surface of the particles. Despite this, it was reported that the dispersion stability of the MRF that the particle composite had been incorporated too and had been increased, and the shear stress of the material had also been increased. Similarly, Zhang et al. (2018) [18] performed a similar synthesis of CIP/polystyrene in their work but with extended characterization in terms of tribology and the rheological properties of the material. It was noted that the yield shear stress, shear viscosity, and the storage modulus the in MRF that contained the polystyrene-coated CIP was lower than that of that contain uncoated CIP, while the tribology tests showed that the former MRF possessed superior wear and frictional properties than the latter.

Meanwhile, a novel poly(glycidyl methacrylate) coating was synthesized by Seung et al. (2019) [29] on the surface of CIP, which later were embedded in both isotropic and anisotropic silicone rubber-based MRE. In this work, MAA, PVP, AIBN and EDGMA were also used as the pre-treatment agent, stabilizer, initiator and cross-linker, respectively, while glycidyl methacrylate (GMA) was used as the monomer. A rougher but bumpy-like coating was observed in this work. According to the team, the dynamic properties of the MRE that consisted of PGMA-coated CIP had superior dynamic properties with a smaller Payne effect than the MRE that contained uncoated CIP.

From these works, it was noted that in most MR materials that adopted the dispersion polymerization for surface coating method, only the monomer was changed for the polymerization to occur. Thus, there is still room for various types of pre-treatment agent, stabilizer, and initiator, as well as the cross-linking agent to be used in this type of method that can be explored.

5. Conclusions

In conclusion, according to the results obtained by studies on particle coating that was then incorporated in MR materials, the coating layer was not only able to protect the magnetic particles from damaging factors such as oxidation, but can also contribute to the enhancement of magnetorheological properties of the material such as the improvement of the sedimentation stability of MRF, increment in wear and friction resistance, as well as reinforcement of the rheological properties of MR materials. It is also emphasized that the coated particles can contribute to more than one advantage that can be offered.

It was clear that almost all MR materials that were impregnated with coated magnetic particles have lower magnetic saturation compared to that of uncoated particles. This is unavoidable because most coatings used are made up from non-magnetic materials such as polymers. However, there were some coated particles that exhibited only a slight drop in magnetic properties, where the materials that were grafted onto the particles were made up from magnetizable materials such as multi-walled carbon nanotubes (MWCNT) [109]. However, when this MWCNT-coated CIP was re-coated with COOH–MWCNT, the magnetic saturation of the particles decreased by about 25%. Therefore, it was ascertained that the thicker the coating layer grafted onto the magnetic particles, the lower the magnetization of the particles, regardless of the type of the coating material.

On the other hand, ascertaining the variety of coating methods that were employed in this MR field may also contribute to the different coating functionalities and properties that can influence the characteristic of the MR materials, with various types of polymers that can be used as the coating layer. Table 3 shows the aforementioned coating methods along with their respective responding variables that are crucial in MR studies in terms of storage modulus, loss factor/loss modulus, MR effect, oxidation stability, sedimentation stability and magnetic saturation as well as other notable outcomes. Some characterizations are unable to be presented into numerical data due to different methods employed by respective authors. Therefore, we only indicated the improvement of the particle coated MR materials characteristics using symbols of ↑ (increase) and ↓ (decrease) when comparing the materials with the uncoated ones. From Table 3, it can be deduced that ATRP is the most commonly used coating method employed in particle coating for MR materials. This might be due to controllable polymerization of ATRP with a wider variety of functionalities that the method could offer, as well as the fact that most MR materials employed in this method have a much lower decrease in magnetic saturation compared to other methods listed in the table below.

Although there are some unavoidable disadvantages when utilizing particle coating in MR materials, such as the lower magnetic saturation, there are still a lot of opportunities for enhancement that can be developed in the near future.

Table 3. Summary of coating functionalities, classified into coating methods as discussed in the text.

Coating Methods	Ref.	Storage Modulus	Loss Modulus/Factor	MR Effect	Chemical Stability	Thermal Stability	Sedimentation Stability	Magnetic Saturation	Other Characteristics
ATRP	[92]	↑	↑	↑			↑	7.02–9.47 emu/g ↓	
	[76]			↑	↑	↑			
	[47]			↑		↑			
	[70]	↓	↑	↑	↑	↑		6.2% ↓ (~10 emu/g)	↑ hydrophobicity ↑ magnetostriction
	[20]			↓	↑			9–14 emu/g ↓	↑ cytotoxicity ↓ yield stress
	[80]	↓	↑	↑	↑	↑		~20 emu/g ↓	↑ roughness
COP	[50]	↑	↓			TGA: ~90wt% at 800 °C ↑	↑	59 emu/g ↓	
	[110]		↓ (strain sweep) ↑ (frequency sweep)					42 emu/g ↓	No comparison for other parameters
	[69]	↑		↓				31.5 emu/g ↓	↑ hydrophobicity
Dispersion polymerization	[29]	↑	↓					45 emu/g ↓	
	[18]	↓					↑	36 emu/g ↓	↓ yield stress ↓ shear viscosity ↑ friction resistance
	[57]				↑		↑		↓ yield stress ↓ shear stress ↑ material removal depth
	[49]							10 emu/g ↓	↑ dispersion stability ≈ shear stress
	[27]						↑	24 emu/g ↓	↓ shear stress ↓ shear viscosity

Author Contributions: Conceptualization, S.K.M.J., N.A.N., U. and S.A.M.; methodology, S.K.M.J.; formal analysis, S.K.M.J.; investigation, S.K.M.J.; resources, N.A.N. and S.A.M.; data curation, S.K.M.J.; writing—original draft preparation, S.K.M.J.; writing—review and editing, N.A.N., U., S.A.A.A., S.A.M. and N.N.; visualization, S.K.M.J.; supervision, N.A.N. and S.A.M.; project administration, S.A.M.; funding acquisition, N.A.N. All authors have read and agreed to the published version of the manuscript.

Funding: This research was funded by the Universiti Teknologi Malaysia—Collaborative Research Grant UTM-CRG, grant number 4B461 and Fundamental Research Grant Scheme FRGS, grant number 5F001.

Acknowledgments: The authors acknowledge the partial financial support provided by Universitas Sebelas Maret for Hibah Kolaborasi Internasional 2021.

Conflicts of Interest: The authors declare no conflict of interest.

References

1. Ladipo, I.L.; Fadly, J.; Faris, W.F. Characterization of Magnetorheological Elastomer (MRE) Engine Mounts. *Mater. Today Proc.* **2016**, *3*, 411–418. [CrossRef]
2. Wang, D.M.; Hou, Y.F.; Tian, Z.Z. A novel high-torque magnetorheological brake with a water cooling method for heat dissipation. *Smart Mater. Struct.* **2013**, *22*, 025019. [CrossRef]
3. Karakoc, K.; Park, E.J.; Suleman, A. Design considerations for an automotive magnetorheological brake. *Mechatronics* **2008**, *18*, 434–447. [CrossRef]
4. Nguyen, P.-B.; Do, X.-P.; Jeon, J.; Choi, S.-B.; Liu, Y.D.; Choi, H.J. Brake performance of core–shell structured carbonyl iron/silica based magnetorheological suspension. *J. Magn. Magn. Mater.* **2014**, *367*, 69–74. [CrossRef]
5. Chen, D.; Yu, M.; Zhu, M.; Qi, S.; Fu, J. Carbonyl iron powder surface modification of magnetorheological elastomers for vibration absorbing application. *Smart Mater. Struct.* **2016**, *25*, 115005. [CrossRef]
6. Deng, H.-X.; Gong, X.-L.; Wang, L.-H. Development of an adaptive tuned vibration absorber with magnetorheological elastomer. *Smart Mater. Struct.* **2006**, *15*, N111–N116. [CrossRef]
7. Deng, H.-X.; Gong, X. Application of magnetorheological elastomer to vibration absorber. *Commun. Nonlinear Sci. Numer. Simul.* **2008**, *13*, 1938–1947. [CrossRef]
8. Du, G.; Chen, X. MEMS magnetometer based on magnetorheological elastomer. *Measurement* **2012**, *45*, 54–58. [CrossRef]
9. Ge, L.; Gong, X.; Wang, Y.; Xuan, S. The conductive three dimensional topological structure enhanced magnetorheological elastomer towards a strain sensor. *Compos. Sci. Technol.* **2016**, *135*, 92–99. [CrossRef]
10. Hu, T.; Xuan, S.; Ding, L.; Gong, X. Stretchable and magneto-sensitive strain sensor based on silver nanowire-polyurethane sponge enhanced magnetorheological elastomer. *Mater. Des.* **2018**, *156*, 528–537. [CrossRef]
11. Skalski, P.; Kalita, K. Implementation of magnetorheological elastomers in transport. *Trans. Inst. Aviat.* **2016**, *245*, 189–198. [CrossRef]
12. Atabay, E.; Ozkol, I. Application of a magnetorheological damper modeled using the current–dependent Bouc–Wen model for shimmy suppression in a torsional nose landing gear with and without freeplay. *J. Vib. Control.* **2014**, *20*, 1622–1644. [CrossRef]
13. Yang, J.; Sun, S.-S.; Tian, T.; Li, W.; Du, H.; Alici, G.; Nakano, M. Development of a novel multi-layer MRE isolator for suppression of building vibrations under seismic events. *Mech. Syst. Signal. Process.* **2016**, *70*, 811–820. [CrossRef]
14. Coyle, S.; Majidi, C.; LeDuc, P.R.; Hsia, K.J. Bio-inspired soft robotics: Material selection, actuation, and design. *Extreme Mech. Lett.* **2018**, *22*, 51–59. [CrossRef]
15. Park, J.; Yoon, G.-H.; Kang, J.-W.; Choi, S.-B. Design and control of a prosthetic leg for above-knee amputees operated in semi-active and active modes. *Smart Mater. Struct.* **2016**, *25*, 085009. [CrossRef]
16. Rabinow, J. The Magnetic Fluid Clutch. *AIEE Trans.* **1948**, *67*, 1308–1315.
17. Plachy, T.; Kutalkova, E.; Sedlacik, M.; Vesel, A.; Masar, M.; Kuritka, I. Impact of corrosion process of carbonyl iron particles on magnetorheological behavior of their suspensions. *J. Ind. Eng. Chem.* **2018**, *66*, 362–369. [CrossRef]
18. Zhang, P.; Dong, Y.Z.; Choi, H.J.; Lee, C.-H. Tribological and rheological tests of core-shell typed carbonyl iron/polystyrene particle-based magnetorheological fluid. *J. Ind. Eng. Chem.* **2018**, *68*, 342–349. [CrossRef]
19. Lee, J.H.; Choi, H.J. Synthesis of core-shell formed carbonyl iron/polydiphenylamine particles and their rheological response under applied magnetic fields. *Colloid Polym. Sci.* **2018**, *296*, 1857–1865. [CrossRef]

20. Cvek, M.; Mrlik, M.; Mosnacek, J.; Babayan, V.A.; Kuceková, Z.; Humpolíček, P.; Pavlinek, V. The chemical stability and cytotoxicity of carbonyl iron particles grafted with poly(glycidyl methacrylate) and the magnetorheological activity of their suspensions. *RSC Adv.* **2015**, *5*, 72816–72824. [CrossRef]
21. Tian, T.; Nakano, M. Fabrication and characterisation of anisotropic magnetorheological elastomer with 45° iron particle alignment at various silicone oil concentrations. *J. Intell. Mater. Syst. Struct.* **2018**, *29*, 151–159. [CrossRef]
22. Liu, J.; Wang, X.; Tang, X.; Hong, R.; Wang, Y.; Feng, W. Preparation and characterization of carbonyl iron/strontium hexaferrite magnetorheological fluids. *Particuology* **2015**, *22*, 134–144. [CrossRef]
23. Lanzetta, M.; Iagnemma, K. Gripping by controllable wet adhesion using a magnetorheological fluid. *CIRP Ann.* **2013**, *62*, 21–25. [CrossRef]
24. Bramantya, M.A.; Sawada, T. The influence of magnetic field swept rate on the ultrasonic propagation velocity of magnetorheological fluids. *J. Magn. Magn. Mater.* **2011**, *323*, 1330–1333. [CrossRef]
25. Mazlan, S.A.; Issa, A.; Chowdhury, H.; Olabi, A. Magnetic circuit design for the squeeze mode experiments on magnetorheological fluids. *Mater. Des.* **2009**, *30*, 1985–1993. [CrossRef]
26. Hu, Z.; Yan, H.; Qiu, H.; Zhang, P.; Liu, Q. Friction and wear of magnetorheological fluid under magnetic field. *Wear* **2012**, *278*, 48–52. [CrossRef]
27. Park, B.J.; Kim, M.S.; Choi, H.J. Fabrication and magnetorheological property of core/shell structured magnetic composite particle encapsulated with cross-linked poly(methyl methacrylate). *Mater. Lett.* **2009**, *63*, 2178–2180. [CrossRef]
28. Agirre-Olabide, I.; Kuzhir, P.; Elejabarrieta, M. Linear magneto-viscoelastic model based on magnetic permeability components for anisotropic magnetorheological elastomers. *J. Magn. Magn. Mater.* **2018**, *446*, 155–161. [CrossRef]
29. Kwon, S.H.; An, J.S.; Choi, S.Y.; Chung, K.H.; Choi, H.J. Poly(glycidyl methacrylate) Coated Soft-Magnetic Carbonyl Iron/Silicone Rubber Composite Elastomer and Its Magnetorheology. *Macromol. Res.* **2019**, *27*, 448–453. [CrossRef]
30. Guan, X.; Dong, X.; Ou, J. Magnetostrictive effect of magnetorheological elastomer. *J. Magn. Magn. Mater.* **2008**, *320*, 158–163. [CrossRef]
31. Khairi, M.H.A.; Mazlan, S.A.; Ubaidillah, U.; Ahmad, S.H.; Choi, S.-B.; Aziz, S.A.A.; Yunus, N.A. The field-dependent complex modulus of magnetorheological elastomers consisting of sucrose acetate isobutyrate ester. *J. Intell. Mater. Syst. Struct.* **2017**, *28*, 1993–2004. [CrossRef]
32. Hapipi, N.; Aziz, S.A.A.; Mazlan, S.A.; Choi, S.B.; Mohamad, N.; Khairi, M.H.A.; Fatah, A.Y.A. The field-dependent rheological properties of plate-like carbonyl iron particle-based magnetorheological elastomers. *Results Phys.* **2019**, *12*, 2146–2154. [CrossRef]
33. Agirre-Olabide, I.; Elejabarrieta, M.J. A new magneto-dynamic compression technique for magnetorheological elastomers at high frequencies. *Polym. Test.* **2018**, *66*, 114–121. [CrossRef]
34. Jung, H.S.; Kwon, S.H.; Choi, H.J.; Jung, J.H.; Gil Kim, Y. Magnetic carbonyl iron/natural rubber composite elastomer and its magnetorheology. *Compos. Struct.* **2016**, *136*, 106–112. [CrossRef]
35. Khimi, S.; Syamsinar, S.; Najwa, T. Effect of Carbon Black on Self-healing Efficiency of Natural Rubber. *Mater. Today Proc.* **2019**, *17*, 1064–1071. [CrossRef]
36. Yu, M.; Zhu, M.; Fu, J.; Yang, P.A.; Qi, S. A dimorphic magnetorheological elastomer incorporated with Fe nano-flakes modified carbonyl iron particles: Preparation and characterization. *Smart Mater. Struct.* **2015**, *24*, 115021. [CrossRef]
37. Bica, I.; Anitas, E.M.; Averis, L.M.E.; Kwon, S.H.; Choi, H.J. Magnetostrictive and viscoelastic characteristics of polyurethane-based magnetorheological elastomer. *J. Ind. Eng. Chem.* **2019**, *73*, 128–133. [CrossRef]
38. Bandarian, M.; Shojaei, A.; Rashidi, A.M. Thermal, mechanical and acoustic damping properties of flexible open-cell polyurethane/multi-walled carbon nanotube foams: Effect of surface functionality of nanotubes. *Polym. Int.* **2011**, *60*, 475–482. [CrossRef]
39. Sung, G.; Kim, J.W.; Kim, J.H. Fabrication of polyurethane composite foams with magnesium hydroxide filler for improved sound absorption. *J. Ind. Eng. Chem.* **2016**, *44*, 99–104. [CrossRef]
40. Baferani, A.H.; Katbab, A.; Ohadi, A. The role of sonication time upon acoustic wave absorption efficiency, microstructure, and viscoelastic behavior of flexible polyurethane/CNT nanocomposite foam. *Eur. Polym. J.* **2017**, *90*, 383–391. [CrossRef]

41. Muhazeli, N.S.; Nordin, N.A.; Mazlan, S.A.; Rizuan, N.; Aziz, S.A.A.; Fatah, A.Y.A.; Ibrahim, Z.; Ubaidillah, U.; Choi, S.-B. Characterization of morphological and rheological properties of rigid magnetorheological foams via in situ fabrication method. *J. Mater. Sci.* **2019**, *54*, 13821–13833. [CrossRef]
42. Xu, J.; Wang, P.; Pang, H.; Wang, Y.; Wu, J.; Xuan, S.; Gong, X. The dynamic mechanical properties of magnetorheological plastomers under high strain rate. *Compos. Sci. Technol.* **2018**, *159*, 50–58. [CrossRef]
43. Zhao, W.; Pang, H.; Gong, X. Novel Magnetorheological Plastomer Filled with NdFeB Particles: Preparation, Characterization, and Magnetic–Mechanic Coupling Properties. *Ind. Eng. Chem. Res.* **2017**, *56*, 8857–8863. [CrossRef]
44. Hapipi, N.M.; Mazlan, S.A.; Ubaidillah, U.; Aziz, S.A.A.; Khairi, M.H.A.; Nordin, N.A.; Nazmi, N. Solvent Dependence of the Rheological Properties in Hydrogel Magnetorheological Plastomer. *Int. J. Mol. Sci.* **2020**, *21*, 1793. [CrossRef]
45. Wu, J.; Gong, X.; Fan, Y.; Xia, H. Physically crosslinked poly(vinyl alcohol) hydrogels with magnetic field controlled modulus. *Soft Matter* **2011**, *7*, 6205–6212. [CrossRef]
46. Qi, S.; Fu, J.; Xie, Y.; Li, Y.; Gan, R.; Yu, M. Versatile magnetorheological plastomer with 3D printability, switchable mechanics, shape memory, and self-healing capacity. *Compos. Sci. Technol.* **2019**, *183*, 107817. [CrossRef]
47. Sutrisno, J.; Fuchs, A.; Sahin, H.; Gordaninejad, F. Surface coated iron particles via atom transfer radical polymerization for thermal-oxidatively stable high viscosity magnetorheological fluid. *J. Appl. Polym. Sci.* **2013**, *128*, 470–480. [CrossRef]
48. Hu, B.; Fuchs, A.; Huseyin, S.; Gordaninejad, F.; Evrensel, C. Atom transfer radical polymerized MR fluids. *Polymer* **2006**, *47*, 7653–7663. [CrossRef]
49. Quan, X.; Chuah, W.; Seo, Y.; Choi, H.J. Core-Shell Structured Polystyrene Coated Carbonyl Iron Microspheres and their Magnetorheology. *IEEE Trans. Magn.* **2014**, *50*, 1–4. [CrossRef]
50. Min, T.H.; Choi, H.J.; Kim, N.-H.; Park, K.; You, C.-Y. Effects of surface treatment on magnetic carbonyl iron/polyaniline microspheres and their magnetorheological study. *Colloids Surf. A Physicochem. Eng. Asp.* **2017**, *531*, 48–55. [CrossRef]
51. Salzman, S.; Romanofsky, H.J.; Jacobs, S.D.; Lambropoulos, J.C. Surface–texture evolution of different chemical-vapor-deposited zinc sulfide flats polished with various magnetorheological fluids. *Precis. Eng.* **2016**, *43*, 257–261. [CrossRef]
52. Saraswathamma, K.; Jha, S.; Rao, P.V. Rheological behaviour of Magnetorheological polishing fluid for Si polishing. *Mater. Today Proc.* **2017**, *4*, 1478–1491. [CrossRef]
53. Paswan, S.K.; Bedi, T.S.; Singh, A.K. Modeling and simulation of surface roughness in magnetorheological fluid based honing process. *Wear* **2017**, *376–377*, 1207–1221. [CrossRef]
54. Lee, J.-W.; Hong, K.-P.; Kwon, S.H.; Choi, H.J.; Cho, M.-W. Suspension Rheology and Magnetorheological Finishing Characteristics of Biopolymer-Coated Carbonyliron Particles. *Ind. Eng. Chem. Res.* **2017**, *56*, 2416–2424. [CrossRef]
55. Zhang, P.; Lee, K.; Lee, C.-H. Fretting friction and wear characteristics of magnetorheological fluid under different magnetic field strengths. *J. Magn. Magn. Mater.* **2017**, *421*, 13–18. [CrossRef]
56. Bombard, A.J.F.; De Vicente, J. Boundary lubrication of magnetorheological fluids in PTFE/steel point contacts. *Wear* **2012**, *296*, 484–490. [CrossRef]
57. Lee, J.W.; Hong, K.P.; Cho, M.W.; Kwon, S.H.; Choi, H.J. Polishing characteristics of optical glass using PMMA-coated carbonyl-iron-based magnetorheological fluid. *Smart Mater. Struct.* **2015**, *24*, 065002. [CrossRef]
58. Hong, K.P.; Song, K.H.; Cho, M.-W.; Kwon, S.H.; Choi, H.J. Magnetorheological properties and polishing characteristics of silica-coated carbonyl iron magnetorheological fluid. *J. Intell. Mater. Syst. Struct.* **2018**, *29*, 137–146. [CrossRef]
59. Xu, Z.; Wang, Q.; Zhu, K.; Jiang, S.; Wu, H.; Yi, L. Preparation and characterization of magnetorheological elastic polishing composites. *J. Intell. Mater. Syst. Struct.* **2019**, *30*, 1481–1492. [CrossRef]
60. Yavas, D.; Yu, T.; Bastawros, A.F. Experimental Feasibility Study of Tunable-Stiffness Polishing Wheel via Integration of Magneto-Rheological Elastomers. In *Mechanics of Composite and Multi-Functional Materials*; Singh, R., Slipher, G., Eds.; Springer International Publishing: Cham, Switzerland, 2020; Volume 5, pp. 85–91.
61. Lokander, M.; Reitberger, T.; Stenberg, B. Oxidation of natural rubber-based magnetorheological elastomers. *Polym. Degrad. Stab.* **2004**, *86*, 467–471. [CrossRef]

62. Ulicny, J.C.; Balogh, M.P.; Potter, N.M.; Waldo, R.A. Magnetorheological fluid durability test—Iron analysis. *Mater. Sci. Eng. A* **2007**, *443*, 16–24. [CrossRef]
63. Burhannuddin, N.L.; Nordin, N.A.; Mazlan, S.A.; Aziz, S.A.A.; Choi, S.-B.; Kuwano, N.; Nazmi, N.; Johari, N. Effects of corrosion rate of the magnetic particles on the field-dependentmaterial characteristics of silicone based magnetorheological elastomers. *Smart Mater. Struct.* **2020**, *29*, 087003. [CrossRef]
64. Utami, D.; Mazlan, S.A.; Imaduddin, F.; Nordin, N.A.; Bahiuddin, I.; Aziz, S.A.A.; Mohamad, N.; Choi, S.-B. Material Characterization of a Magnetorheological Fluid Subjected to Long-Term Operation in Damper. *Materials* **2018**, *11*, 2195. [CrossRef] [PubMed]
65. Shafrir, S.N.; Romanofsky, H.J.; Skarlinski, M.; Wang, M.; Miao, C. Zirconia Coated Carbonyl Iron Particle-Based Magnetorheological Fluid for Polishing. *Appl. Opt.* **2009**, *48*, 6797–6810. [CrossRef] [PubMed]
66. Behrooz, M.; Sutrisno, J.; Zhang, L.; Fuchs, A.; Gordaninejad, F. Behavior of magnetorheological elastomers with coated particles. *Smart Mater. Struct.* **2015**, *24*, 35026. [CrossRef]
67. Perales-Martínez, I.A.; Palacios-Pineda, L.M.; Lozano-Sánchez, L.M.; Martínez-Romero, O.; Puente-Cordova, J.G.; Elías-Zúñiga, A. Enhancement of a magnetorheological PDMS elastomer with carbonyl iron particles. *Polym. Test.* **2017**, *57*, 78–86. [CrossRef]
68. Sedlacik, M.; Mrlik, M.; Babayan, V.; Pavlinek, V. Magnetorheological elastomers with efficient electromagnetic shielding. *Compos. Struct.* **2016**, *135*, 199–204. [CrossRef]
69. Yu, M.; Qi, S.; Fu, J.; Zhu, M.; Chen, D. Understanding the reinforcing behaviors of polyaniline-modified carbonyl iron particles in magnetorheological elastomer based on polyurethane/epoxy resin IPNs matrix. *Compos. Sci. Technol.* **2017**, *139*, 36–46. [CrossRef]
70. Cvek, M.; Mrlík, M.; Ilcíková, M.; Mosnáček, J.; Münster, L.; Pavlínek, V. Synthesis of Silicone Elastomers Containing Silyl-Based Polymer-Grafted Carbonyl Iron Particles: An Efficient Way to Improve Magnetorheological, Damping, and Sensing Performances. *Macromolecules* **2017**, *50*, 2189–2200. [CrossRef]
71. Wang, J.-S.; Matyjaszewski, K. "Living"/Controlled Radical Polymerization. Transition-Metal-Catalyzed Atom Transfer Radical Polymerization in the Presence of a Conventional Radical Initiator. *Macromolecules* **1995**, *28*, 7572–7573. [CrossRef]
72. Matyjaszewski, K.; Xia, J. Atom Transfer Radical Polymerization. *Chem. Rev.* **2001**, *101*, 2921–2990. [CrossRef] [PubMed]
73. Król, P.; Chmielarz, P. Recent advances in ATRP methods in relation to the synthesis of copolymer coating materials. *Prog. Org. Coat.* **2014**, *77*, 913–948. [CrossRef]
74. Matyjaszewski, K. *Overview: Fundamentals of Controlled/Living Radical Polymerization*; American Chemical Society: Washington, DC, USA, 1998; pp. 2–30.
75. Morick, J.; Buback, M.; Matyjaszewski, K. Activation—Deactivation Equilibrium of Atom Transfer Radical Polymerization of Styrene up to High Pressure. *Macromol. Chem. Phys.* **2011**, *212*, 2423–2428. [CrossRef]
76. Fuchs, A.; Sutrisno, J.; Gordaninejad, F.; Caglar, M.B.; Yanming, L. Surface polymerization of iron particles for magnetorheological elastomers. *J. Appl. Polym. Sci.* **2010**, *117*, 934–942. [CrossRef]
77. Li, Y.; Li, Q.; Zhang, C.; Cai, P.; Bai, N.; Xu, X. Intelligent self-healing superhydrophobic modification of cotton fabrics via surface-initiated ARGET ATRP of styrene. *Chem. Eng. J.* **2017**, *323*, 134–142. [CrossRef]
78. Parida, D.; Serra, C.A.; Gómez, R.I.; Garg, D.K.; Hoarau, Y.; Bouquey, M.; Muller, R. Atom Transfer Radical Polymerization in Continuous Microflow: Effect of Process Parameters. *J. Flow Chem.* **2014**, *4*, 92–96. [CrossRef]
79. Monge, S.; Darcos, V.; Haddleton, D.M. Effect of DMSO used as solvent in copper mediated living radical polymerization. *J. Polym. Sci. Part. A Polym. Chem.* **2004**, *42*, 6299–6308. [CrossRef]
80. Cvek, M.; Mrlik, M.; Sevcik, J.; Sedlacik, M. Tailoring Performance, Damping, and Surface Properties of Magnetorheological Elastomers via Particle-Grafting Technology. *Polymers* **2018**, *10*, 1411. [CrossRef]
81. Bondarev, D.; Borská, K.; Šoral, M.; Moravčíková, D.; Mosnáček, J. Simple tertiary amines as promotors in oxygen tolerant photochemically induced ATRP of acrylates. *Polymer* **2019**, *161*, 122–127. [CrossRef]
82. Khezri, K.; Fazli, Y. Evaluation of the effect of hydrophobically modified silica aerogel on the ARGET ATRP of styrene and butyl acrylate. *Microporous Mesoporous Mater.* **2019**, *280*, 236–242. [CrossRef]
83. Zaborniak, I.; Chmielarz, P.; Martinez, M.R.; Wolski, K.; Wang, Z.; Matyjaszewski, K. Synthesis of high molecular weight poly(n-butyl acrylate) macromolecules via seATRP: From polymer stars to molecular bottlebrushes. *Eur. Polym. J.* **2020**, *126*, 109566. [CrossRef]

84. Trevisanello, E.; De Bon, F.; Daniel, G.; Lorandi, F.; Durante, C.; Isse, A.A.; Gennaro, A. Electrochemically mediated atom transfer radical polymerization of acrylonitrile and poly(acrylonitrile-b-butyl acrylate) copolymer as a precursor for N-doped mesoporous carbons. *Electrochim. Acta* **2018**, *285*, 344–354. [CrossRef]
85. Wang, G.; Wang, Z.; Lee, B.; Yuan, R.; Lu, Z.; Yan, J.; Pan, X.; Song, Y.; Bockstaller, M.R.; Matyjaszewski, K. Polymerization-induced self-assembly of acrylonitrile via ICAR ATRP. *Polymer* **2017**, *129*, 57–67. [CrossRef]
86. Chmielarz, P.; Park, S.; Simakova, A.; Matyjaszewski, K. Electrochemically mediated ATRP of acrylamides in water. *Polymer* **2015**, *60*, 302–307. [CrossRef]
87. Ma, Y.; Cai, M.; He, L.; Luo, X. Enhanced protein retention on poly(caprolactone) via surface initiated polymerization of acrylamide. *Appl. Surf. Sci.* **2016**, *360*, 20–27. [CrossRef]
88. Matyjaszewski, K.; Coca, S.; Gaynor, S.G.; Wei, M.; Woodworth, B.E. Controlled radical polymerisation in the presence of oxygen. *Macromolecules* **1998**, *31*, 5967–5969. [CrossRef]
89. Horn, M.; Matyjaszewski, K. Solvent Effects on the Activation Rate Constant in Atom Transfer Radical Polymerization. *Macromolecules* **2013**, *46*, 3350–3357. [CrossRef]
90. París, R.; De La Fuente, J.L. Solvent effect on the atom transfer radical polymerization of allyl methacrylate. *J. Polym. Sci. Part. A Polym. Chem.* **2005**, *43*, 6247–6261. [CrossRef]
91. Zhu, K.; Zhu, Z.; Zhou, H.; Zhang, J.; Liu, S. Precisely installing gold nanoparticles at the core/shell interface of micellar assemblies of triblock copolymers. *Chin. Chem. Lett.* **2017**, *28*, 1276–1284. [CrossRef]
92. Cvek, M.; Mrlik, M.; Ilcíková, M.; Plachy, T.; Sedlacik, M.; Mosnacek, J.; Pavlinek, V. A facile controllable coating of carbonyl iron particles with poly(glycidyl methacrylate): A tool for adjusting MR response and stability properties. *J. Mater. Chem. C* **2015**, *3*, 4646–4656. [CrossRef]
93. Eyiler, E.; Walters, K.B. Magnetic iron oxide nanoparticles grafted with poly(itaconic acid)-block-poly(N-isopropylacrylamide). *Colloids Surf. A Physicochem. Eng. Asp.* **2014**, *444*, 321–325. [CrossRef]
94. Sapurina, I.Y.; Shishov, M.A. Oxidative Polymerization of Aniline: Molecular Synthesis of Polyaniline and the Formation of Supramolecular Structures. *New Polym. Spec. Appl.* **2012**, *740*, 272. [CrossRef]
95. Higashimura, H.; Kobayashi, S. Oxidative Polymerization. In *Encyclopedia of Polymer Science and Technology*; Wiley-Interscience: New York, NY, USA, 2016; pp. 1–37. [CrossRef]
96. Jamari, S.K.M.; Kasi, R.; Ismail, L.; Nor, N.A.M.; Subramanian, R.R.; Subramaniam, T.R.; Vengadaesvaran, B.; Arof, A.K.M. Studies on anticorrosion properties of polyaniline-TiO2 blended with acrylic-silicone coating using electrochemical impedance spectroscopy. *Pigment. Resin Technol.* **2016**, *45*, 18–23. [CrossRef]
97. Kalendova, A.; Veselý, D.; Stejskal, J. Organic coatings containing polyaniline and inorganic pigments as corrosion inhibitors. *Prog. Org. Coat.* **2008**, *62*, 105–116. [CrossRef]
98. He, Z.; Fang, Y.; Wang, X.; Pang, H. Microwave absorption properties of PANI/CIP/Fe$_3$O$_4$ composites. *Synth. Met.* **2011**, *161*, 420–425. [CrossRef]
99. Fan, M.; He, Z.; Pang, H. Microwave absorption enhancement of CIP/PANI composites. *Synth. Met.* **2013**, *166*, 1–6. [CrossRef]
100. Lin, K.-Y.; Hu, L.-W.; Chen, K.-L.; Siao, M.-D.; Ji, W.-F.; Yang, C.-C.; Yeh, J.-M.; Chiu, K.-C. Characterization of polyaniline synthesized from chemical oxidative polymerization at various polymerization temperatures. *Eur. Polym. J.* **2017**, *88*, 311–319. [CrossRef]
101. Amer, I.; Young, D.A. Chemically oxidative polymerization of aromatic diamines: The first use of aluminium-triflate as a co-catalyst. *Polymer* **2013**, *54*, 505–512. [CrossRef]
102. Kong, H.; Song, J.; Jang, J. One-step fabrication of magnetic γ-Fe2O3/polyrhodanine nanoparticles using in situ chemical oxidation polymerization and their antibacterial properties. *Chem. Commun.* **2010**, *46*, 6735. [CrossRef]
103. Yuan, G.; Wang, H. Facile synthesis and performance of polypyrrole-coated sulfur nanocomposite as cathode materials for lithium/sulfur batteries. *J. Energy Chem.* **2014**, *23*, 657–661. [CrossRef]
104. Bai, R.; Zhang, X. Polypyrrole-Coated Granules for Humic Acid Removal. *J. Colloid Interface Sci.* **2001**, *243*, 52–60. [CrossRef]
105. Liu, J.; Wang, Q.; Liu, P. Poly(1,5-diaminoanthraquinone) coated carbon cloth composites as flexible electrode with extraordinary cycling stability for symmetric solid-state supercapacitors. *J. Colloid Interface Sci.* **2019**, *546*, 60–69. [CrossRef] [PubMed]
106. Wang, Q.; Li, R.R.; Zhou, X.Z.; Li, J.; Lei, Z. Polythiophene-coated nano-silicon composite anodes with enhanced performance for lithium-ion batteries. *J. Solid State Electrochem.* **2016**, *20*, 1331–1336. [CrossRef]

107. Slomkowski, S.; Alemán, J.V.; Gilbert, R.G.; Hess, M.; Horie, K.; Jones, R.G.; Kubisa, P.; Meisel, I.; Mormann, W.; Penczek, S.; et al. Terminology of polymers and polymerization processes in dispersed systems (IUPAC Recommendations 2011). *Pure Appl. Chem.* **2011**, *83*, 2229–2259. [CrossRef]
108. Kawaguchi, S.; Ito, K. Dispersion polymerization. In *Polymer Particles*; Springer: Berlin/Heidelberg, Germany, 2005; Volume 175, pp. 299–328.
109. Fang, F.F.; Liu, Y.D.; Choi, H.J. Carbon nanotube coated magnetic carbonyl iron microspheres prepared by solvent casting method and their magneto-responsive characteristics. *Colloids Surf. A Physicochem. Eng. Asp.* **2012**, *412*, 47–56. [CrossRef]
110. Kim, J.H.; Fang, F.F.; Choi, H.J.; Seo, Y. Magnetic composites of conducting polyaniline/nano-sized magnetite and their magnetorheology. *Mater. Lett.* **2008**, *62*, 2897–2899. [CrossRef]

Publisher's Note: MDPI stays neutral with regard to jurisdictional claims in published maps and institutional affiliations.

© 2020 by the authors. Licensee MDPI, Basel, Switzerland. This article is an open access article distributed under the terms and conditions of the Creative Commons Attribution (CC BY) license (http://creativecommons.org/licenses/by/4.0/).

MDPI
St. Alban-Anlage 66
4052 Basel
Switzerland
Tel. +41 61 683 77 34
Fax +41 61 302 89 18
www.mdpi.com

Materials Editorial Office
E-mail: materials@mdpi.com
www.mdpi.com/journal/materials

www.ingramcontent.com/pod-product-compliance
Lightning Source LLC
LaVergne TN
LVHW070602100526
838202LV00012B/538